TA645 CAS

STRUCTURAL ANALYSIS

STRUCTURAL ANALYSIS
The Solution of Statically Indeterminate Structures

by

W. FISHER CASSIE,

Ph.D. (*St. Andrews*), M.S. (*Illinois*), F.R.S.E.

Member Institution of Civil Engineers
Member Institution of Structural Engineers
Associate Member Town Planning Institute
Member American Concrete Institute

Professor of Civil Engineering
University of Durham

WITH A FOREWORD BY
H. JOHN COLLINS, M.C., M.Sc.

Chadwick Professor of Civil and
Municipal Engineering, University
of London. President of the Institution
of Structural Engineers 1946-47

LONGMANS, GREEN AND CO.
LONDON • NEW YORK • TORONTO

LONGMANS, GREEN AND CO LTD
6 & 7 CLIFFORD STREET LONDON W I

ALSO AT MELBOURNE AND CAPE TOWN

LONGMANS, GREEN AND CO INC
55 FIFTH AVENUE NEW YORK 3

LONGMANS, GREEN AND CO
215 VICTORIA STREET TORONTO I

ORIENT LONGMANS LTD
BOMBAY CALCUTTA MADRAS

FIRST PUBLISHED   ...   1947
New Impression   ...   1948
*New Impression (with
minor corrections)*   ...   1951

Printed in Great Britain by William Clowes & Sons, Limited, London and Beccles

FOREWORD

The subject of statically indeterminate structures has frequently been treated in English engineering text-books by application to special cases.

Since the publication in Italy by Castigliano in 1879 of a treatise on the principle of least work and its application to determining the stresses in a structure containing superfluous bars, many authors have expounded and illustrated the general principle that the work stored in an elastic system in stable equilibrium is always the smallest possible. Alternative methods for determining the stresses in structural members have also been evolved, notably those methods developed by Maxwell.

Of the numerous works that have been written upon the subject of statically indeterminate structures many are excellent examples of mathematical gymnastics rather than of engineering application.

In the following pages the Author has successfully endeavoured to supply a want which is felt by many engineers and students of engineering, by bringing together in one volume applications of the general principles and their illustration by worked solutions. He has worked out numerous examples showing the practical application of theory to problem in logical sequence and with ample detail in order to assist the reader. The steps of the reader are guided in paths often trodden by and therefore familiar to the Author, who is thereby enabled frequently to recommend a straight course without wastage of time spent by the student in search for a route.

The Author has drawn liberally from his store of knowledge, a store stocked with experience gained in the teaching of engineering over many years, and the reader cannot fail to appreciate not only his knowledge of the subject but also his ability to impart that knowledge.

Nam non solum scire aliquid, artis est, sed quaedam ars etiam docendi.

Many years of happy association with Professor Cassie have strengthened the writer's conviction that not only does he know his subject, but he also possesses in no small measure "a certain art in teaching it."

H. JOHN COLLINS.

University College,
London.

AIMS AND WORKING SUGGESTIONS

This book is particularly directed to engineering students who are in the last two years of a University or Technical College course or who are studying with a view to becoming Associate Members of one of the engineering Institutions.

A student who has reached this stage usually has no serious difficulty in reading and comprehending the mathematical background to the Theory of Structures, but when using his theoretical knowledge in the solution of particular problems, his work is often clumsy and inefficient, or may even be completely at fault.

This volume, then, assumes that the theoretical basis for the solution of statically indeterminate structures is understood or readily available, and concentrates on numerical problems. Its detailed aims are—

(1) to show, by means of series of graded examples, the use of the best known methods of solving statically indeterminate structures;

(2) to illustrate the relative applicability and ease of operation of the various methods by means of exercises which afford direct comparisons of the work of different chapters;

(3) to compare (in view of the increasing use and importance of continuous structures) the accuracy of the more recently developed methods of solution with that of the "exact" methods;

(4) to encourage the reader to master the subject, regardless of the requirements of examinations, by leading him to the solution of more involved problems than he is likely to encounter in his student work.

It has been assumed that the reader is familiar with the methods of determining direct forces, shearing forces, bending moments, deflections and influence lines for statically determinate structures, and that he can carry out simple differentiation and integration.

The problems are intended to illustrate, in the simplest possible way, the resolution of difficulties commonly encountered by students. They are not intended to be "practical", but one or two problems have been inserted with a view to showing how the theoretical methods may be linked up to design problems.

Each problem takes the student a further step towards the solution of the more complex frames, and the examples should thus be studied in the order in which they appear in the chapter concerned. Each chapter, however, may be read independently of the others, but a good knowledge of the facts of Chapters 1 and 2 is of considerable assistance in dealing with more advanced methods.

It is important to note that, whether the student reads the book in order to learn fully the application of an unfamiliar method of calculation or merely to determine the best way of solving a particular problem, *it is essential that he should work out each problem for himself.* No mere reading of the book will result in an intimate understanding of how to overcome the perplexities involved.

The average student usually encounters his chief difficulty in evolving a technique of writing down the succeeding steps of a solution in the most efficient manner, and the only way to master this technique is by repeated practice. For this reason, each chapter deals with the same types of relatively simple problems. Unusual and "tricky" examples have been avoided, for the basic steps in each method are better understood by the solution of many similar straightforward problems, than by the study of exceptional or ingenious solutions.

In order to establish a thorough, basic knowledge of the methods described, *all problems and exercises should be worked out afresh without reference to the text,* and it should be remembered that not only the student's final result but also his method of attack should be submitted to criticism and to comparison with the text. Only in this way can improvement in speed and neatness be obtained. Some problems have been left uncompleted in order to encourage personal work of this kind.

The Author is indebted to R. W. Holmes, B.Sc., A.M.Inst.C.E. for his skilful preparation of the diagrams, to F. Pickering, A.M.I.Struct.E., and J. A. Williams, A.M.Inst.C.E., for their detailed checking of much of the calculation, and to A. M. Ballantyne, Ph.D., A.M.Inst.C.E., for the help given in reading proofs. Several classes of senior civil engineering students of this College also gave valuable assistance in the preparation of the text and the Institution of Structural Engineers kindly permitted the reprinting of the matter in Chapter 7.

King's College,   W. F. C.
Newcastle.
January, 1947.

CONTENTS

CHAPTER 1

DEFINITION OF STATICALLY INDETERMINATE STRUCTURES

The objective to which the work of analysing any structure is directed is that of determining, for every portion of the structure, the direct, shearing and bending stresses. When these are known, the members of the structure may be suitably proportioned to resist the external loading.

The problem must always be solved in two stages:

(1) The applied external load must be balanced by reactions and "fixing" moments, acting in defined directions on the structure, the latter being considered as a rigid body of known shape and size, but not necessarily of any specific construction.

(2) Since the forces and bending moments applied at various sections of the structure are known from (1), the balancing forces and moments called into play within the structure itself can now be determined, and the members composing the structure made sufficiently strong to resist the induced stresses.

Stages (1) and (2) are carried out in two different ways according as the structure is (a) statically determinate or (b) statically indeterminate. It is with the statically indeterminate structure that this book is concerned, and it is important that the student should be able to draw a clear distinction between (a) and (b), and to appreciate their fundamental difference, before proceeding to study methods of analysis.

STATICALLY DETERMINATE STRUCTURES

In structures of this type, the external reactions of stage (1) and the internal forces of stage (2) can be completely determined by equating the known to the unknown forces (and moments) according to the laws of statics. 1,0 (a) to (e) shows a number of statically determinate structures. The supporting forces and moments and the internal forces and moments are just sufficient in number to

1

balance the external loading. If one of the supports of (a) or (c), for example, were to fail, or one of the members of (e) to be removed, the structure in question would collapse. The student is expected to be thoroughly familiar with the solution of such structures.

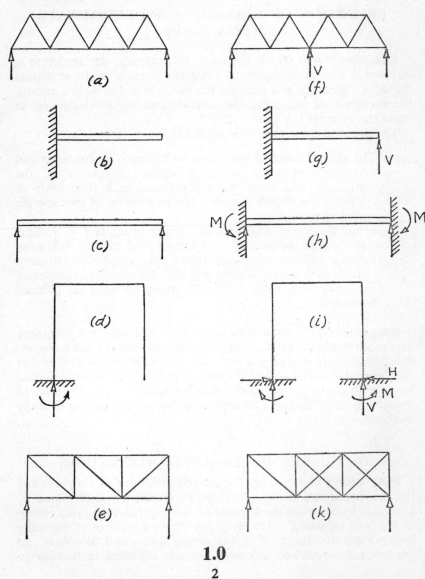

STATICALLY INDETERMINATE STRUCTURES

If structures (**a**) to (**d**) are supplied with additional supports or if structure (**e**) possesses more internal members than are sufficient to prevent the frame from collapsing, the structures become statically indeterminate, as is shown from (**f**) to (**k**).

Every additional force or moment applied in this way must first be evaluated before the methods of statics can be used to make a complete determination of stresses.

The methods employed vary, as will be realised in the following chapters, but all utilise, in some way, the deformation or displacement of the structure under its external loading.

If n forces and/or moments must be released to convert a statically indeterminate structure to the statically determinate condition, the structure is statically indeterminate in the nth degree, and n equations involving the n unknowns must be written down from consideration of the deformations or displacements of the structure.

For (**f**) and (**g**) only one such equation is required; (**h**) is statically indeterminate to the second degree, since two moments must be removed before the condition in (**c**) is reached, and (**i**) requires three equations to determine H, V and M.

DIFFERENCES BETWEEN STATICALLY DETERMINATE AND STATICALLY INDETERMINATE STRUCTURES

Statically Determinate	*Statically Indeterminate*
Cross - sectional areas and Moments of Inertia of beams and frame members need not be known in order to find the internal forces in the structure.	Cross-sectional areas and Moments of Inertia of beams and frame members have an important bearing on the values of the internal forces and moments, and must be determined before the investigation is begun.
Slight settlement of a support, or slight lack of fit of a frame member, has no effect on the internal forces and moments.	Slight settlement of a support, or slight lack of fit of a frame member, is an important factor in determining the internal forces and moments, and such settlement or lack of fit must be estimated before the investigation is begun.

CHAPTER 2

METHOD OF AREA MOMENTS

The simplest method of analysing the stresses in a bent beam, which is statically indeterminate to a low degree, is that of Area Moments. This method forms a good introduction to methods of wider application.

Principle of Area-moments (Fig. 2.01)

   (a) In any portion, AB, of a bent beam, the angle, ϕ, between the tangents to the beam at A and B is numerically equal to the area of the M/EI diagram between these points.

   (b) In any portion, AB, of a bent beam, the displacement of A from the tangent to the beam at B is equal to the moment of the area of the M/EI diagram between A and B, *taken about A.*

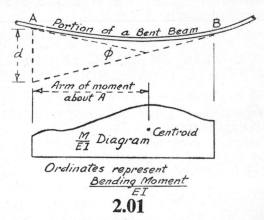

Ordinates represent
Bending Moment
EI

2.01

Since the displacement, d, is always very small in relation to the length of the beam, it is immaterial whether it is assumed to be measured at right angles to the beam or at right angles to the tangent.

Two points of importance should be noted and remembered. When calculating displacements, take the moment of the M/EI area *about the point where the displacement is required.* The figures found by following the principle (b), above, *do not necessarily indicate the deflection of the beam from its original position.*

4

Procedure

(1) Remove the statically indeterminate forces and moments so that the structure is left in a statically determinate condition (Fig. **1.0**).

(2) Apply the external loading and draw the Bending Moment Diagram. This is often known as the "free" bending moment diagram.

(3) Divide each ordinate of the bending moment diagram by the term EI, and draw the M/EI diagram.

(4) Remove the external loading from the beam and apply the statically indeterminate forces and moments. Draw the M/EI diagram for these forces and moments only.

(5) From the conditions of support the nett slope or deflection of the beam at one or more sections is usually known. Find the slope and/or deflection at these sections from (3) in terms of the known loading, and from (4) in terms of the unknown reactions or moments, and, by comparison, determine the "unknowns."

Convention of Signs.

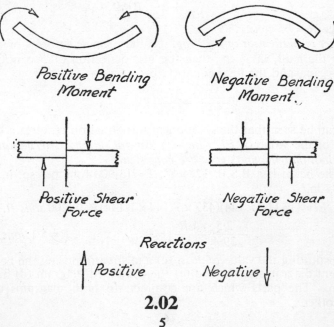

Positive Bending Moment

Negative Bending Moment.

Positive Shear Force

Negative Shear Force

Reactions

Positive   Negative

2.1 Since this is a cantilever, the tangent at A, when the prop is removed (in accordance with the procedure outlined above) is horizontal. Thus the "displacement of B from the tangent at A" is also the vertical deflection of B from its original position. E and I are constant over the whole length of the beam, and their values need not be known.

(1) Remove the redundant reaction R. B deflects through d, a distance equal to the moment of the area of the M/EI diagram (due to the loading) about B. **(b)**

$$\text{Deflection of } B = d = -\frac{1}{EI}\left(\frac{1}{3} \times 100 \times 10\right) \times \frac{3}{4} \times 10 = -\frac{2500}{EI} \text{ } ft.$$

(2) Remove the loading and replace R. B deflects upward through a distance d' equal to the moment of the M/EI diagram (due to R) about B. **(c)**

$$\text{Deflection of } B = d' = +\frac{1}{EI}\left(\frac{1}{2} \times 10R \times 10\right) \times \frac{2}{3} \times 10 = +\frac{1000}{3EI} \text{ } R \text{ } ft.$$

(3) Since, with a rigid, unyielding support, there is actually no deflection of B, d and d' must be numerically equal.

$$-\frac{2500}{EI} + \frac{1000}{3EI}R = 0,$$

whence $R = 7 \cdot 5$ *tons*.

Effect of Settlement of the Prop. If the prop R deforms or settles under the load, say $\frac{1}{4}$ in., then the nett deflection of B is not zero, but $-\frac{1}{4}$ in., and the final equation becomes

$$-\frac{2500}{EI} + \frac{1000}{3EI}R = -\frac{1}{48}.$$

It can be seen that the solution of this equation requires a knowledge of the value of EI. *Be very careful, in any determination of this kind, that the units used are consistent.*

If the beam is a B.S.B. $12'' \times 8''$, $E = 13,000$ tons per sq. in., and $I = 437$ in.[4]

$$EI = (13,000 \times 144)(437 \times 1/144 \times 1/144) = 39,400 \text{ } tons\text{-}ft.^2$$

Hence $\qquad -\dfrac{2 \cdot 5}{39 \cdot 4} + \dfrac{R}{118 \cdot 2} = -\dfrac{1}{48} \qquad\qquad R = 5 \cdot 1 \text{ } tons.$

Substituting the value of R in **(c)** and superimposing the bending moment diagrams of **(b)** and **(c)** the combined diagram **(d)** may be drawn. The parts which are common to both diagrams cancel each other.

2.1

The shape of the deflected form of a frame is important in more complex building frames, and sketches such as (e) should be drawn for all problems.

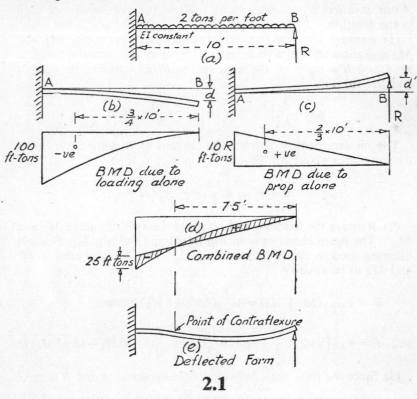

2.1

Exercise

Solve problem 3.1 by *Area Moments*.

7

2.2 The ends of this beam are each supported in a horizontal direction. Both before and after loading, the tangents at A and B are horizontal, the angle between them being zero; neither A nor B suffers any displacement. These two conditions are used in the solution. EI is constant.

(1) Remove the two statically indeterminate moments M_L and M_R and allow the beam to deflect. The angle between the tangents at A and B is equal to the area of the M/EI diagram between A and B. (b) (c)

$$\phi = +\frac{1}{EI}\left(\frac{1}{2} \times 30 \times 16\right) = +\frac{240}{EI} \text{ radians.}$$

The displacement of B from the tangent at A is the moment of the same area about B. (b) (c)

$$d = +\frac{240}{EI} \times \frac{28}{3} = +\frac{2240}{EI} \text{ ft.}$$

(2) Remove the loading, and re-apply the two moments M_L and M_R. The beam bends as is shown in (d) and the bending moment diagram used in the calculations is (e). This time, assuming M_L and M_R to be positive

$$\phi' = +\frac{16}{2EI}(M_L + M_R) = +\frac{1}{EI}(8M_L + 8M_R) \text{ radians}$$

and $d' = +\dfrac{1}{EI}\left(8M_L \times \dfrac{32}{3} + 8M_R \times \dfrac{16}{3}\right) = +\dfrac{1}{EI}(85\cdot33M_L + 42\cdot67M_R)\,ft.$

(3) Since the final angle between the tangents at A and B is zero,

$$+8M_L + 8M_R + 240 = 0$$

and since the final displacement of B from the tangent at A is zero,

$$+85\cdot33M_L + 42\cdot67M_R + 2240 = 0.$$

Solving these equations simultaneously

$$M_L = -22\cdot5 \text{ ft.-tons.} \qquad M_R = -7\cdot5 \text{ ft.-tons.}$$

This application of the Method of Area Moments will be found useful in determining the preliminary values of the fixing moments when using the Method of Moment Distribution of Chapter 6.

Draw the combined bending moment diagram and the deflected form of the beam. (f) (g)

8

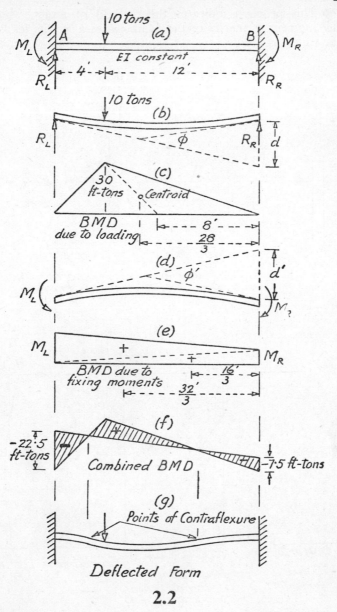

(a) 10 tons M_L A EI constant B M_R R_L 4' 12' R_R

(b) 10 tons R_L ϕ R_R d

(c) 30 ft-tons Centroid BMD due to loading 8' $\frac{28}{3}$

(d) M_L ϕ' d' $M_?$

(e) M_L + + M_R BMD due to fixing moments $\frac{16}{3}$ $\frac{32}{3}$

(f) -22.5 ft-tons Combined BMD -7.5 ft-tons

(g) Points of Contraflexure Deflected Form

2.3 This problem indicates the method of attack when the moment of inertia of the beam varies along its length. It can be stated as follows:

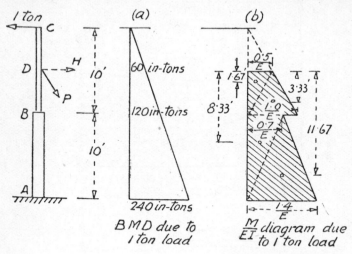

(a) BMD due to 1 ton load

(b) $\frac{M}{EI}$ diagram due to 1 ton load

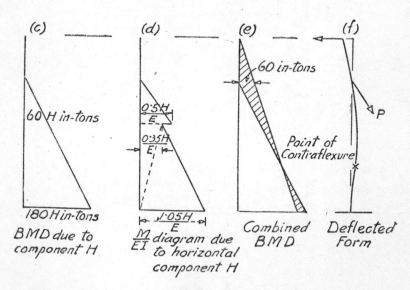

(c) BMD due to component H

(d) $\frac{M}{EI}$ diagram due to horizontal component H

(e) Combined BMD

(f) Deflected Form

2.3

A hollow steel pole is vertical and 20 ft. high. It is made in two sections of equal length, the lower of which has a moment of inertia equal to 171 in.⁴ and the upper, a moment of inertia of 120 in.⁴ The horizontal pull of 1 ton in a cable attached to the top of the pole is counteracted by a stay fastened to the pole at D, a point 5 ft. from the top. Find the tension in the stay, which makes an angle of 35° with the vertical, if the position of D, relative to A, remains unchanged after loading.

(1) Remove the stay and allow the 1 ton load to act. (a) is the bending moment diagram, and (b) is the M/EI diagram, each ordinate of (a) having been divided by 171 in.⁴ in the lower portion, and 120 in.⁴ in the upper portion. The value of E is constant and need not be known.

Displacement of D from the tangent at A is equal to the moment of the shaded area about D. (b)

Displacement of D:

$$d = \frac{144}{E}(1\cdot4 \times 5 \times 11\cdot67 + 0\cdot7 \times 5 \times 8\cdot33 + 2\cdot5 \times 3\cdot33 + 1\cdot25 \times 1\cdot67)$$

$$= \frac{144 \times 121\cdot3}{E} \ in.$$

(2) Remove the 1 ton load and re-apply the pull P of the stay. We are concerned with only the horizontal component of P, since this component (H) causes the bending moment in the pole, the vertical component causing only a vertical direct stress in the pole. (c) is the bending moment diagram in these circumstances and (d) is the M/EI diagram.

Displacement of D:

$$d' = -\frac{144H}{E}(1\cdot05 \times 5 \times 11\cdot67 + 0\cdot35 \times 5 \times 8\cdot33 + 0\cdot5 \times 2\cdot5 \times 3\cdot33)$$

$$= -\frac{144 \times 80H}{E} \ in.$$

(3) D does not, finally, deflect from its original position, therefore,

$$121\cdot3 - 80H = 0,$$

whence $H = 1\cdot52$ tons, and P, the load in the tie $= \dfrac{1\cdot52}{\sin 35°} = 2\cdot65$ *tons.*

The combined bending moment diagram is drawn by superimposing (a) and (c), the point X representing a point of contraflexure or change in the direction of bending of the beam.

2.4 *Two cantilevers, one vertically above the other, are connected at the ends by a rod 4 ft. in length. Find the maximum bending moment in each of the cantilevers, and the force in the tie rod.*

Since both cantilevers are of the same length and cross-section, and there is a heavier load on the lower one, the rod AB is in tension.

The deflection of A, therefore, is due to the effect of the 2-ton load, and the downward pull of the rod AB.

The deflection of B is due to the effect of the two 4-ton loads and the upward pull of the rod AB.

$E = 13,000$ *tons per sq. in.* $\qquad\qquad$ $I = 1000$ *in.*[4]

Cross-sectional area of rod $AB = 0.25$ *sq. in.*

$EI = 13,000 \times 144 \times 1000 \times 1/144 \times 1/144 = 90,250$ *tons-ft.*^{2}

Upper Cantilever. The moment of the area of the diagram (**a**) about A = deflection of A

$$= -\frac{1}{2EI} \times 9 \times 4\cdot5 \times 7\cdot5 = -\frac{152}{EI} \, ft.$$

The further downward deflection of A due to the pull in AB (**d**)

$$= -\frac{1}{2EI} \times 9P \times 9 \times 6 = -\frac{243P}{EI} \, ft.$$

Total deflection of A

$$= -\frac{1}{EI}(152 + 243P) \, ft.$$

Lower Cantilever. Moment of the areas of the two diagrams (**b**) and (**c**) about B = deflection of B, downwards

$$= -\frac{1}{2EI}(24 \times 6 \times 7 + 12 \times 3 \times 8) = -\frac{648}{EI} \, ft.$$

The upward deflection of B due to the pull in AB (**d**)

$$= +\frac{1}{2EI} \times 9P \times 9 \times 6 = +\frac{243P}{EI} \, ft.$$

Total deflection of B

$$= \frac{1}{EI}(-648 + 243P) \, ft.$$

Rod AB. The extension of AB under the unknown load P

$$= \frac{Pl}{AE} = -\frac{P \times 48}{0\cdot25 \times 13,000} = -0\cdot0147P \, in.$$

Combined Effect. If AB were non-extensible then the deflections of A and B would be equal. The deflection of B, however, must

12

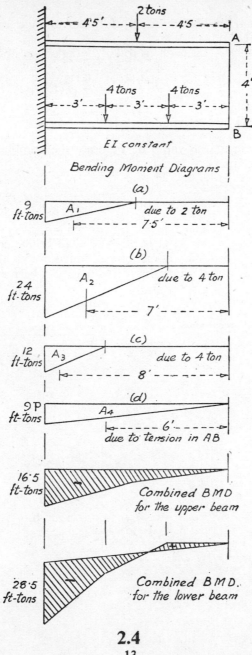

2 tons

4·5' ←→ 4·5

A

4'

4 tons   4 tons

3'   3'   3'

B

EI constant

Bending Moment Diagrams

(a)

9 ft-tons   A₁   due to 2 ton

7·5'

(b)

24 ft-tons   A₂   due to 4 ton

7'

(c)

12 ft-tons   A₃   due to 4 ton

8'

(d)

9 P ft-tons   A₄

6'

due to tension in AB

16·5 ft-tons   Combined BMD for the upper beam

28·5 ft-tons   Combined BMD for the lower beam

be greater than the deflection of A by an amount equal to the extension of the rod AB.

Deflection of B—extension of rod AB=deflection of A

$$\frac{1}{EI}(-648+243P)+\frac{0\cdot0147P}{12}=-\frac{1}{EI}(152+243P)$$

whence $P=0\cdot83$ *tons* tension.

Having obtained the value of P, draw the combined bending moment diagrams.

2.5 A girder of this kind is often used in bridge construction, and the bending moments for which the beam must be designed can be found when the reaction R_B has been evaluated.

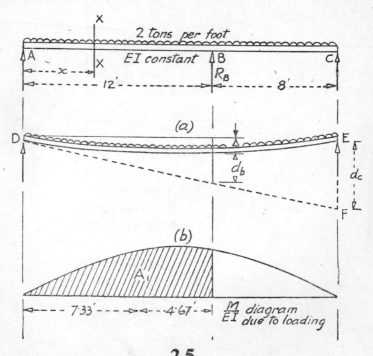

(1) **Remove** the redundant reaction at B and allow the beam to deflect. (a)

The displacement of B from the tangent at A is d_b and is equal to the moment of the shaded portion of diagram (b) about B. The best method of obtaining this area and its moment about A or B is by integration.

The ordinate of the $\dfrac{M}{EI}$ diagram $= +\dfrac{1}{EI}(20x - x^2)$.

The shaded area $= \dfrac{1}{EI}\displaystyle\int_0^{12} M\,dx = +\dfrac{1}{EI}\int_0^{12}(20x - x^2)\,dx = +\dfrac{864}{EI}$.

The moment of the shaded area of the diagram taken about A

$$= \dfrac{1}{EI}\int_0^{12}(20x - x^2)x\,dx = +\dfrac{6336}{EI}\,ft.$$

Distance to centroid of shaded area

$$\text{from } A = \dfrac{6336}{864} = 7\cdot33\,ft.$$

$$\text{from } B = 12 - 7\cdot33 = 4\cdot67\,ft.$$

Therefore, the displacement of B from the tangent at A

$$d_b = +\dfrac{864}{EI} \times 4\cdot67 = +\dfrac{4034}{EI}\,ft.$$

The displacement of C from the tangent at $A = d_c$ and is equal to the moment of the area of the whole M/EI parabola about C. Displacement of C from tangent at A

$$d_c = \dfrac{2}{3} \times \dfrac{100}{EI} \times 20 \times 10 = \dfrac{13{,}330}{EI}\,ft.$$

But the required deflection of the beam from its original position is $\varDelta$

$$\text{and} \qquad d_b + \varDelta = \dfrac{12}{20}d_c \text{ (from triangle } DEF\text{).}$$

$$\therefore \qquad \varDelta = \dfrac{12}{20} \times \dfrac{13{,}330}{EI} - \dfrac{4034}{EI} = \dfrac{3966}{EI}\,ft.$$

15

2.5

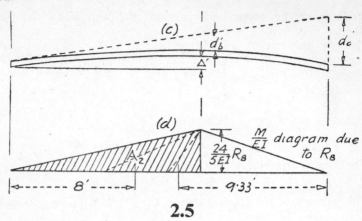

M/EI diagram due to R_B

$$\leftarrow \cdots \cdots 8' \cdots \cdots \rightarrow | \leftarrow \cdots 9\cdot 33' \cdots \cdots \rightarrow$$

2.5

(2) Remove the superimposed load and replace the reaction R_B. The beam deflects upwards. (c) The M/EI diagram is shown in (d).

Displacement of B from the tangent at A :

$$d_b' = \text{shaded triangle} \times 4'$$

$$= -\frac{1}{2} \times \frac{24}{5EI} R_B \times 12 \times 4 = -\frac{115}{EI} R_B \, ft.$$

Displacement of C from the tangent at A :

$$d_c' = \text{moment of whole } M/EI \text{ diagram about } C$$

$$= -\frac{1}{2} \times \frac{24}{5EI} R_B \times 20 \times 9\cdot 33 = -\frac{448}{EI} R_B.$$

As above, $d_b' + \Delta' = \frac{12}{20} d_c'$.

$$\therefore \Delta' = -\frac{1}{EI} \left(\frac{3 \times 448}{5} R_B + 115 R_B \right) = -\frac{154}{EI} R_B.$$

But if there is no settlement of R_B, $\Delta + \Delta' = 0$

$$\therefore R_B = \frac{3966}{154} = +25\cdot 8 \, tons.$$

By taking moments about A and C it is now possible to determine the reactions at C and A, and, by superposition of (b) and (d) the combined B.M.D. may be drawn.

16

THEOREM OF THREE MOMENTS

The Theorem of Three Moments is derived directly from the Area-moment principle, and this is probably its most useful application.

This theorem gives the relationship between the moments at the supports of a continuous beam and the loading on the beam. Referring to Fig. 2.03, which shows a prismatic beam of uniform section,

$$M_A l_1 + 2M_B(l_1 + l_2) + M_C l_2 = -\frac{6A_1 \bar{x}_1}{l_1} - \frac{6A_2 \bar{x}_2}{l_2}.$$

A_1 and A_2 are the areas of the "free" bending moment diagram for each span; i.e. the B.M.D. caused by the loading if each span were independent and freely supported at the ends. M_A, M_B and M_C represent the unknown bending moment at each of the supports; these are often called the "support moments". $\bar{x}$ is the distance from the centroid of each bending moment area to the *outside* support of the span considered.

The equation or equations determined in this way give the values of the support moments and consequently the bending moment diagram can be drawn. The reactions at each support and the shear force diagram can then be obtained by the methods of statics, but in this connection the convention of signs is of importance. It is in determining the reactions that the student is likely to make a mistake, and particular attention should be paid to Fig. 2.02 which shows the convention used in this chapter.

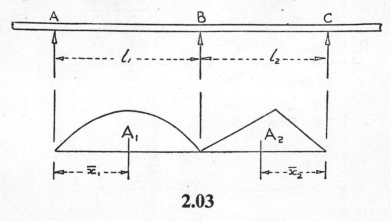

2.03

2.6 This problem is similar to Problem **2.5**, but instead of considering R_B as the redundancy, the value of M_B is found from the "three-moment" equation and the evaluation of the other reactions and moments is then easily accomplished by the methods of statics.

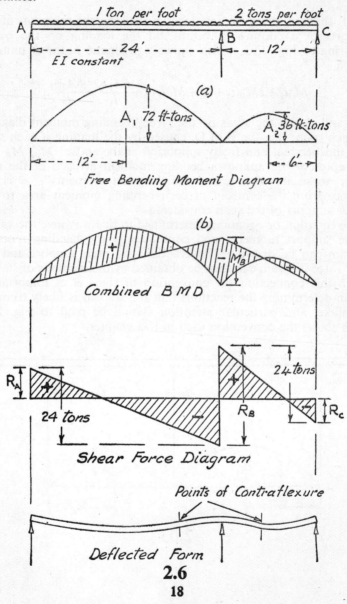

(a)

Free Bending Moment Diagram

(b)

Combined BMD

Shear Force Diagram

Points of Contraflexure

Deflected Form

2.6

Free Bending Moment Areas. Evaluate the areas of the free bending moment diagram for each span. **(a)**

$$A_1 = \frac{2}{3} \times \frac{w_1 l_1^2}{8} \times l_1 = \frac{w_1 l_1^3}{12} = 1152 \ tons\text{-}ft.^2$$

$$A_2 = \frac{w_2 l_2^3}{12} = 288 \ tons\text{-}ft.^2$$

When the areas have been determined, find the values of $\bar{x}$ for each span, remembering that these lengths must be measured to the *outside* support of the span considered. In this problem, since the B.M.D.'s are symmetrical, $\bar{x}$ is 12 ft. for A_1 and 6 ft. for A_2.

Equations. Substitute all known values in the " three-moment" equation, noting that M_A and M_C are zero, since the beam is simply supported at A and C.

$$M_A l_1 + 2M_B(l_1 + l_2) + M_C l_2 = -\left(\frac{6A_1 \bar{x}_1}{l_1} + \frac{6A_2 \bar{x}_2}{l_2}\right)$$
$$0 + 72M_B + 0 = -3(1152 + 288)$$
$$M_B = -60 \ ft.\text{-}tons.$$

The complete bending moment diagram for the whole span can now be drawn as is shown in **(b)**.

Reactions. In finding the reactions the student should pay particular attention to the convention of signs, and plus signs should be written down as well as minus signs when the equations are being formed.

Take moments about B of all forces to the left of B, remembering that clockwise moments to the left of the section are positive.

$$+R_A l_1 - \frac{w_1 l_1^2}{2} = \text{moment at } B$$

$$+24R_A - \frac{24^2}{2} = -60$$

$$R_A = +9 \cdot 5 \ tons \ (upward).$$

Take moments about C of all forces to the left of C

$$+R_A(l_1 + l_2) + R_B l_2 - w_1 l_1 \left(l_2 + \frac{l_1}{2}\right) - \frac{w_2 l_2^2}{2} = \text{moment at } C$$

$$+36R_A + 12R_B - 24 \times 24 - 144 = 0$$

$$R_B = +31 \cdot 5 \ tons \ (upward).$$

From these results, the shear force diagram may be drawn as is shown, R_C being $= 48 - (R_A + R_B) = +7 \cdot 0 \ tons \ (upward)$.

If a continuous beam of this type rests on more than three supports, there is more than one unknown support moment. In such circumstances the "three-moment" equation must be written

19

2.6

down several times, each equation referring to two consecutive spans in the manner just described.

For instance, if a continuous girder rests on five supports, A, B, C, D and E, all at the same level, the equation must be written for A–B–C, B–C–D and C–D–E. These three equations, solved simultaneously, give the support moments at B, C and D.

In such a problem a larger number of areas must be dealt with and the equations solved simultaneously, but such an extension of Problem **2.6** follows the same steps of "Areas", "Equations" and "Reactions."

Exercises

Solve, by *Area Moments*, problems 3.6, 5.1, 5.5, 7.1 and 7.2.

Solve, by *Theorem of Three Moments*, problems 3.1, 3.2, 3.6, 5.1, 6.1, 6.2, 6.3, 6.4, 7.1, 7.2.

2.7 In this beam one end is fixed in a horizontal direction, and the fixing moment at that end is consequently unknown. A second equation is therefore required for the solution of the beam.

The moment at C may be imagined to be developed by spans CD and DE identical in dimensions and loading to spans CB and BA, and forming a continuation of ABC towards the right.

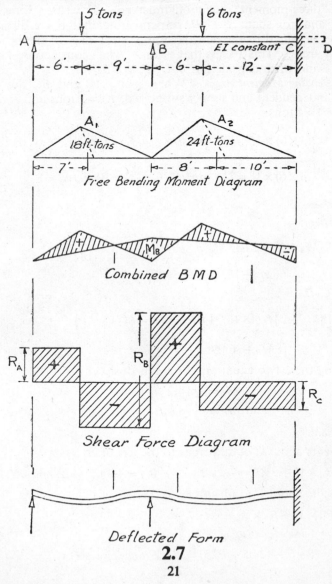

Free Bending Moment Diagram

Combined B M D

Shear Force Diagram

Deflected Form

2.7

A more convenient method, however, is to imagine one extra span *CD* which is sufficiently heavily loaded to develop the required moment at C. The moment at *C* can be developed by this method, in an infinite number of ways and, in the limit, a beam *CD* may be assumed of zero length and carrying an infinite load.

The "three-moment" equation must now be written for spans *A–B–C* and for spans *B–C–D*, remembering that the $\bar{x}$ distances must be measured to the outside supports for each of the bending moment areas. This means that $\bar{x}$ for A_2 when considering *A–B–C* is 10 ft., and when considering *B–C–D*, is 8 ft.

Free Bending Moment Areas. When spans *AB* and *BC* are considered independent and simply supported, $R_A = 3$ tons and R_B (for span *BC*)$= 4$ tons.

$$A_1 = \frac{1}{2}(3 \times 6) \times 15 = 135 \; tons\text{-}ft.^2 \qquad \bar{x}_1 = 7 \; ft.$$

$$A_2 = \frac{1}{2}(4 \times 6) \times 18 = 216 \; tons\text{-}ft.^2 \qquad \bar{x}_2 = 10 \; ft., \; \bar{x}_2' = 8 \; ft.$$

Equations.

A–B–C

$$15M_A + 66M_B + 18M_C = -6\left(\frac{135 \times 7}{15} + \frac{216 \times 10}{18}\right)$$

$$66M_B + 18M_C = -1098.$$

B–C–D

$$18M_B + 2M_C(18+0) + 0 \times M_D = -6\left(\frac{216 \times 8}{18}\right)$$

$$18M_B + 36M_C = -576.$$

Solving these two equations simultaneously,

$$M_B = -14 \cdot 2 \; ft.\text{-}tons$$
$$\text{and } M_C = - \; 8 \cdot 9 \; ft.\text{-}tons.$$

Reactions.

Moments about *B* of all forces to the left of *B*

$$+R_A \times 15 - 5 \times 9 = -14 \cdot 2. \qquad R_A = +2 \cdot 05 \; tons \; (upward).$$

Moments about *C* of all forces to the left of *C*

$$+R_A \times 33 + R_B \times 18 - 5 \times 27 - 6 \times 12 = -8 \cdot 9.$$

$$R_B = +7 \cdot 25 \; tons \; (upward).$$

$$R_C = (5+6) - (2 \cdot 05 + 7 \cdot 25) = +1 \cdot 7 \; tons \; (upward).$$

2.8 This problem illustrates a fact which is not always appreciated by students when solving "three-moment" problems, namely that the moment at the end of a loaded cantilever is known, and need not be treated as statically indeterminate. By the convention of signs, this moment is negative, and the minus sign must appear in the relevant equation.

Free Bending Moment Areas.

$$A_1 = \frac{2}{3} \times 25 \times 10 = \frac{500}{3} \ tons\text{-}ft.^2 \qquad \bar{x}_1 = 5 \ ft.$$

$$A_2 = 7\cdot5 \times \frac{8}{2} = 30 \ tons\text{-}ft.^2 \qquad \bar{x}_2 = \frac{11}{3} \ ft., \quad \bar{x}_2' = \frac{13}{3} \ ft.$$

Equations.

A–B–C
($M_A = -4 \ ft.\text{-}tons.$)

$$-4l_1 + 2M_B(l_1 + l_2) + M_C l_2 = -\frac{6 \times 2500}{30} - \frac{6 \times 30 \times 11}{24}.$$

$$36M_B + 8M_C = -542\cdot5.$$

B–C–D.

$$M_B l_2 + 2M_C(l_2 + 0) = -\frac{6 \times 30 \times 13}{24}.$$

$$8M_B + 16M_C = -97\cdot5.$$

Solving these simultaneously,

$$M_B = -15\cdot43 \ ft.\text{-}tons. \quad M_C = +1\cdot62 \ ft.\text{-}tons.$$

Reactions.
Moments about B of all forces to the left of B

$$+10R_A - 2 \times 12 - 20 \times 5 = -15\cdot43.$$

$$R_A = +10\cdot85 \ tons \ (upward).$$

Moments about C of all forces to the left of C

$$+8R_B + 10\cdot85 \times 18 - 2 \times 20 - 20 \times 13 - 4 \times 3 = +1\cdot62.$$

$$R_B = +14\cdot80 \ tons \ (upward).$$

A common mistake is to include the values of the support moments in the left-hand side of equations such as that just used to find R_B. The support moments such as M_A and M_B are balanced within the beam and have no additional effect on moments at a distance. For instance, there is a moment of -4 ft.-tons on *both* sides of the support A, and a moment of $-15\cdot43$ ft.-tons on *both* sides of support B. M_C, however, is not balanced, but is applied to the beam from without. Such applied or "fixing" moments must be taken into account.

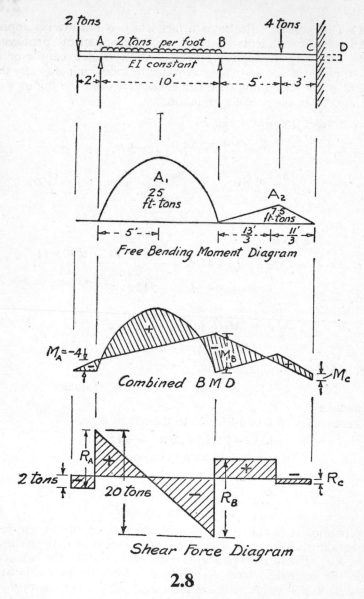

Free Bending Moment Diagram

Combined B M D

Shear Force Diagram

2.8

As an illustration of the above remarks, R_C may be found by taking moments about B or A.

Moments about B of all forces to the right of B.

$$+1·62+8R_C-20=-15·43.$$
$$R_C=+0·35 \; tons \; (upward).$$

Moments about A of all forces to the right of A.

$$+1·62+18R_C+14·8\times10-4\times15-20\times5=-4.$$

The total upward forces should equal the total downward forces: $10·85+14·80+0·35=26$ tons.

2.9 When, as in this instance, a load covers only a portion of the span, the bending moment area is best considered as several separate areas, the area of that portion under the loading being most conveniently found by integration.

This beam is statically indeterminate to the third degree and requires three equations for its solution. As before, an imaginary span of length zero must be added to the left of $A(DA)$ and a similar span to the right of $C(CE)$. The "three-moment" equations must then be written for the spans $D-A-B$, $A-B-C$ and $B-C-E$.

Free Bending Moment Areas.

$$A_1=\int_0^6 Mdx=\int_0^6\left(\left(4·5x-\frac{x^2}{2}\right)\right)dx=45 \; tons\text{-}ft.^2$$

$$A_2=\frac{1}{2}\times9\times6=27 \; tons\text{-}ft.^2 \quad \text{Total area}=45+27=72 \; tons\text{-}ft.^2$$

$$A_1\bar{x}_1=\int_0^6 Mxdx=\int_0^6\left(4·5x^2-\frac{x^3}{2}\right)dx=162 \; tons\text{-}ft.^3$$

Taking moments of both areas about A

$$\text{Moment}=A_1\bar{x}_1+A_2\times8$$
$$=162+27\times8=378. \quad \bar{x}=\frac{378}{45+27}=5·25 \; ft.$$

Therefore, the centroid of the combined area A_1+A_2 is $5·25$ ft. from A.

Another method of obtaining these figures is by drawing the line FG in (**a**). The curved portion of the bending moment diagram above FG has an area of $\frac{2}{3}\times\frac{1\times6^2}{8}\times6=18 \; tons\text{-}ft.^2$ and the moment of this area about $A=54 \; tons\text{-}ft.^3$

The triangular portion of the bending moment diagram (below FG) has an area 54 tons-ft.^{2} and a moment about A of 54×6 $=324$ tons-ft.^{3} This gives a total area of 72 tons-ft.^{2} and a total moment of area about A of 378 tons-ft.^{3} as before.

25

2.9

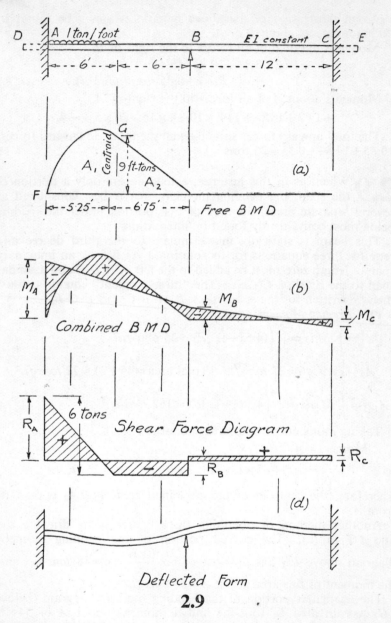

D E┊ A 1 ton/foot   B   EI constant   C ┊E

← 6' → ← 6' → ← 12' →

(a)

G
Centroid
A_1 9 ft-tons
A_2
F
← 5·25' → ← 6·75' →   Free B M D

(b)

M_A −   +
M_B M_C
Combined B M D

R_A 6 tons   Shear Force Diagram
+
− R_B + R_C

(d)

Deflected Form

2.9

Equations.

D–A–B

$$M_D \times 0 + 2M_A \times 12 + M_B \times 12 = -\frac{72 \times 6 \cdot 75}{2}.$$

$$24M_A + 12M_B = -243.$$

A–B–C

$$12M_A + 2M_B \times 24 + 12M_C = -\frac{72 \times 5 \cdot 25}{2}.$$

$$12M_A + 48M_B + 12M_C = -189.$$

B–C–E

$$12M_B + 24M_C = 0.$$

From this last equation it appears that $M_C = -\frac{1}{2}M_B$. This can be put into words as follows:

If a beam BC is fixed in position and direction at C, and a moment is applied at B (the beam being otherwise unloaded) the fixing moment at C is numerically equal to half of the applied moment and is of the opposite sign; provided that the position of B relative to C remains unchanged.

From this statement, which is of value in more advanced investigations, it follows that the point of contraflexure is at a distance from the fixed end equal to one-third of the length of the beam.

The solution of the three equations simultaneously gives

$$M_A = -9 \cdot 1 \; ft.\text{-}tons. \quad M_B = -1 \cdot 9 \; ft.\text{-}tons. \quad M_C = +0 \cdot 95 \; ft.\text{-}tons.$$

Reactions. When taking moments on the left of B and C it is necessary to include the unbalanced fixing moment M_A.

Moments about B of all forces to the left of B.

$$M_A + 12R_A - 6 \times 9 = M_B.$$
$$-9 \cdot 1 + 12R_A - 54 = -1 \cdot 9.$$
$$R_A = +5 \cdot 1 \; tons \; (upward).$$

Moments about C of all forces to the left of C.

$$M_A + 24R_A + 12R_B - 6 \times 21 = M_C.$$
$$-9 \cdot 1 + 5 \cdot 1 \times 24 + 12R_B - 126 = +0 \cdot 95.$$
$$R_B = +1 \cdot 14 \; tons \; (upward).$$

Total upward force $= 6 \cdot 24$ *tons.* Total load on the beam $= 6$ *tons.* Therefore, $R_C = -0 \cdot 24$ *tons (downward).*

From these results the bending moment and shear force diagrams can be drawn.

2.10
The "three-moment" equation used in the preceding problems is a special case of the general "three-moment" equation which takes into account differences in the moments of inertia of the beams forming the two spans.

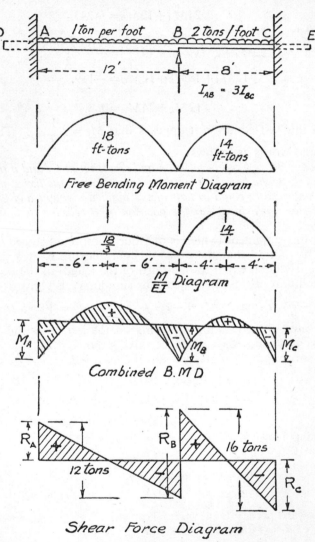

D A *1 ton per foot* B *2 Tons/foot* C E

12'   8'

$I_{AB} = 3I_{BC}$

18 ft-tons   14 ft-tons

Free Bending Moment Diagram

$\frac{18}{3}$ $\frac{14}{1}$

6'   6'   4'   4'

$\frac{M}{EI}$ Diagram

M_A M_B M_C

Combined B.M.D

R_A R_B 16 tons

12 tons R_C

Shear Force Diagram

2.10

28

The general equation reads

$$M_A\frac{l_1}{I_1}+2M_B\left(\frac{l_1}{I_1}+\frac{l_2}{I_2}\right)+M_C\frac{l_2}{I_2}=-\frac{6A_1\bar{x}_1}{l_1}-\frac{6A_2\bar{x}_2}{l_2}.$$

The right-hand side of the equation is apparently unaltered, but it must be remembered that A_1 and A_2 now refer to the M/I diagrams instead of merely the bending moment diagrams as in the preceding problems. E is considered to be constant throughout, and its value need not be known.

The built-in beam of this problem is made up of two portions, AB and BC, whose respective moments of inertia are as $3:1$. The actual values of the moments of inertia are not required.

Free Bending Moment Areas.

Left-hand span $\dfrac{wl^3}{12}=144$ *tons-ft.*^{2}

Right-hand span $\dfrac{wl^3}{12}=85\cdot33$ *tons-ft.*^{2} (See Problem **2.6**.)

Relative M/I Areas.

$$A_1=\frac{144}{3}=48.$$

$$A_2=85\cdot33.$$

Equations.

D–A–B.

$$2M_A\times\frac{12}{3}+\frac{12}{3}M_B=-\frac{6\times48}{2}.$$

$$2M_A+M_B=-36.$$

A–B–C.

$$\frac{12}{3}M_A+2M_B\left(\frac{12}{3}+8\right)+8M_C=-3(48+85\cdot33).$$

$$M_A+6M_B+2M_C=-100.$$

B–C–E.

$$8M_B+2M_C\times8=-\frac{6}{2}\times85\cdot33.$$

$$M_B+2M_C=-32.$$

From these equations

$$M_A=-12\cdot44,\ \ M_B=-11\cdot12,\ \ M_C=-10\cdot44\ \textit{ft.-tons};$$

$$R_A=+6\cdot1,\ \ R_B=14\cdot0,\ \ R_C=+7\cdot9\ \textit{tons};$$

from which the bending moment and shear force diagrams may be drawn.

2.11 When the beam carries a distributed load which varies in intensity along its length, the area of the bending moment or M/EI diagram and the moment of that area about one of the supports, must be found by integration.

In this problem, the beam is built-in at both ends, and carries a distributed load varying from zero at one end to 1 ton per ft. at the other, the variation being of parabolic form.

Following the procedure of Problem **2.2**, the first step is to remove the two redundant moments. In (b) the beam is shown in a statically determinate condition.

The equation to the parabola (origin of co-ordinates at R, and the vertex at V) i

$$w = Cx(40-x)$$

when $w=1$, $x=20$. $\therefore$ $w=\dfrac{x}{400}(40-x)$.

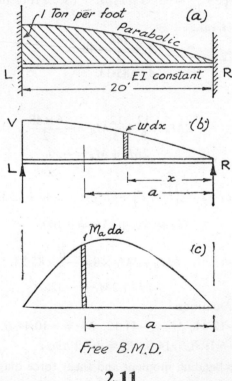

Free B.M.D.

2.11

30

Free Bending Moment Diagram. The first step is to find the "free" reactions. Taking moments about the right-hand end of the beam,

$$20R_L = \int_0^{20} wx\,dx = \frac{1}{400} \int_0^{20} x^2(40-x)\,dx \ tons$$

whence $R_L = 8\cdot3$ *tons*, and similarly, $R_R = 5\cdot0$ *tons*.

The bending moment at a section a from R is

$$M_a = R_R a - \int_0^a w(a-x)\,dx \qquad \text{(substitute for } w\text{)}$$

$$= 5a - \frac{a^3}{1200}\left(20 - \frac{a}{4}\right) ft.\text{-}tons.$$

The area of the bending moment diagram is the integration of the elementary area shown in (c)

$$\int_0^{20} M_a\,da = \int_0^{20} \left\{ 5a - \frac{a^3}{1200}\left(20 - \frac{a}{4}\right) \right\} da = 466\cdot7 \ tons\text{-}ft.^2$$

The moment of the free bending moment diagram about R is given by the integration of the moment of the elementary area, about R.

$$\int_0^{20} M_a a\,da = 4880 \ tons\text{-}ft.^3$$

Equations. With these figures, and remembering that EI is constant, complete the solution by the methods of Problem **2.2**. The result is $M_L = -26\cdot6$ ft.-tons and $M_R = -20\cdot0$ ft.-tons. Draw the bending moment and shearing force diagrams.

If the moment of inertia of the beam had varied continuously along the beam, it would have appeared in the integrations as a function of a.

2.12 An important application of the Theorem of Three Moments is in the analysis of bending moments and shears in a rim-bearing swing bridge. When closed across the river, this type of bridge is carried by four supports (A, B, C, D), one at each bank and two provided by the roller supports on the pier. For the loading shown, the "three-moment" equation must be applied to A–B–C and to B–C–D. For the sake of illustration assume that the effective moment of inertia of BC is twice that of AB.

2.12

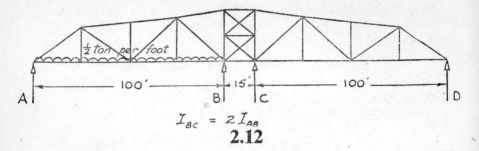

$$I_{BC} = 2I_{AB}$$

2.12

Equations.

A–B–C.

$$100M_A + 2M_B\left(100 + \frac{15}{2}\right) + \frac{15}{2}M_C = -3 \times \frac{10,000}{16} \times \frac{2}{3} \times 100$$

$$215M_B + 7 \cdot 5M_C = -125,000.$$

B–C–D.

$$7 \cdot 5M_B + 2M_C\left(\frac{15}{2} + 100\right) + 100M_D = 0$$

$$M_B = -28 \cdot 7M_C.$$

whence $M_B = -582 \cdot 6$ *ft.-tons*; $M_C = +20 \cdot 3$ *ft.-tons*.

Reactions.

Moments about *B* of all forces to the left of *B*

$$100R_A - 2500 = -582 \cdot 6. \qquad R_A = \quad 19 \cdot 2 \text{ tons.}$$

Moments about *C* of all forces to the left of *C*

$$115R_A - 50 \times 65 + 15R_B = +20 \cdot 3. \qquad R_B = \quad 70 \cdot 9 \text{ tons.}$$

Moments about *C*

from the other end $100R_D = +20 \cdot 3.$ $\qquad R_D = \quad 0 \cdot 2$ *tons.*

Moments about *B* of all forces to the right of *B*

$$115R_D + 15R_C = -582 \cdot 6. \qquad R_C = -40 \cdot 3 \text{ tons } (\textbf{down})$$

$$\text{Sum} = \quad \overline{50 \cdot 0} \text{ tons.}$$

Draw the shear force and bending moment diagrams for the bridge.

CHAPTER 3

METHOD OF STRAIN ENERGY

General Principles

The amount of strain energy or work stored in a loaded structure depends on the magnitude of the direct, shear and bending stresses imposed on the various parts of the structure. In pin-jointed frames, for example, where the members are in tension or compression, the work stored depends on direct forces only. Such frames are discussed in Chapter 9.

In Chapters 2 to 8 the problems deal with beams and with frames having rigid joints. In such structures, direct stress, shear stress and bending stress may all occur at any section, and the total strain energy stored in the beam or frame depends on the magnitudes of the three types of stress. It is generally conceded, however, that the work done by the direct and shear forces is so small in comparison with that done by bending that only the latter need be considered when calculating statically indeterminate reactions or moments.

It must be remembered, however, that although the work done by shear and direct forces may be considered negligible, yet the values of the shear and direct stresses must be included in the final stress values when the strength of the structure is being checked.

In any member of a structure subjected to bending the total internal work or strain energy is

$$\bar{U} = \int_0^l \frac{M^2 ds}{2EI}$$

where M is the bending moment at any point on the member caused by the combined effect of the imposed loads and the supporting forces and moments, whether statically determinate or not. The integration must be taken over the whole length of the member, of which ds is an element of length.

Castigliano showed that the partial differential coefficient of the strain energy in a structure, with respect to a load P acting on the structure, is equivalent to the displacement of P along its line of action.

$$\Delta_P = \frac{\partial \bar{U}}{\partial P} = \int_0^l \frac{2M}{2EI} \frac{\partial M}{\partial P} ds = \int_0^l \frac{M}{EI} \cdot \frac{\partial M}{\partial P} ds.$$

3.1

Similarly, the partial differential coefficient of the strain energy in a structure with respect to a moment acting on the structure is equivalent to the angle through which that portion of the structure rotates when the moment is applied.

$$\phi = \frac{\partial \overline{U}}{\partial M_x} = \int_0^l \frac{M}{EI} \cdot \frac{\partial M}{\partial M_x} ds.$$

It very often occurs that the forces P and the moments M_x are the supporting forces and moments of a statically indeterminate structure, and if the supports of the structure do not give way under the action of the loading, then there is no deformation of the structure at the points of support and the expressions just quoted can be equated to zero. If the differential coefficient of the strain energy is thus zero, then the strain energy itself is a minimum. The term "Method of Least Work" may therefore be found applied to Strain Energy determinations.

BEAMS AND FRAMES HAVING ONE REDUNDANT REACTION

3.1 If the support R were removed, the structure would become a cantilever, and it is therefore statically indeterminate to the first degree. A single equation, in addition to those of statics, is sufficient to determine the stresses in the beam.

From the principle of Strain Energy the equation required is

$$\frac{I}{EI} \int_0^l M \frac{\partial M}{\partial R} dx = 0.$$

EI is a constant whose value is not required and which can be cancelled from subsequent calculations.

The integration must be made in two sections, since the bending moment expression changes at half span. The work is most conveniently done in tabular form. Working from the right-hand end of the beam, the table is as follows:

	Bending Moment (M)	$\partial M/\partial R$	Limits
Right half . . .	Rx	$+x$	0 to 5
Left half . . .	$Rx - \frac{w}{2}(x-5)^2$	$+x$	5 to 10

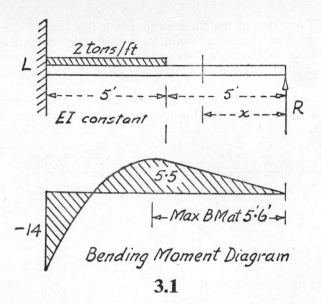

5·5

|← Max BM at 5·6 →|

−14

Bending Moment Diagram

3.1

It will be found easier to work each integration separately, than to try to evaluate all terms in one operation. By this means there is less chance of a mistake being made in the evaluation of the definite integrals.

$$\int_0^5 M\frac{\partial M}{\partial R}dx = \int_0^5 (Rx)x\,dx = 41\cdot67R.$$

$$\int_5^{10} M\frac{\partial M}{\partial R}dx = \int_5^{10}\{Rx-(x-5)^2\}x\,dx = 291\cdot7R - 364\cdot6.$$

Summing and equating to zero,

$$\int_0^{10} M\frac{\partial M}{\partial R}dx = 333R - 364\cdot6 = 0.$$

$$R = +1\cdot1 \ tons.$$

The bending moment diagram can now be drawn by substituting this value of R in the "bending moment" column of the table. Shear force at any point (and, if necessary, internal stresses) may now be found by the methods of Chapter 2.

3.2 If any one of the three supports of this beam were to be removed, the beam would be statically determinate. Only one equation is therefore required to find one of the reactions. Thereafter the methods of statics can be used to determine the others.

By taking moments, two of the reactions are first expressed in terms of the third, so that only one unknown appears in the equation.

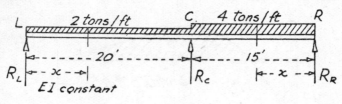

3.2

Taking moments about R

$$35R_L + 15R_C = 1450. \quad \therefore R_L = 41 \cdot 4 - 0 \cdot 43R_C.$$

Taking moments about L

$$35R_R + 20R_C = 2050. \quad R_R = 58 \cdot 6 - 0 \cdot 57R_C.$$

It must be remembered that the integration must be taken over the whole length of the beam, and in this problem it is easier to work from both ends towards the centre than from one end only. EI is again constant over the whole length.

Member	Bending Moment	Moment re-written	$\partial M/\partial R_C$	Limits
RC	$R_R x - 2x^2$	$58 \cdot 6x - 0 \cdot 57xR_C - 2x^2$	$-0 \cdot 57x$	0 to 15
LC	$R_L x - x^2$	$41 \cdot 4x - 0 \cdot 43xR_C - x^2$	$-0 \cdot 43x$	0 to 20

$$\int_0^{15} M \frac{\partial M}{\partial R_C} dx = \int_0^{15} (58 \cdot 6x - 0 \cdot 57xR_C - 2x^2)(-0 \cdot 57x) dx$$
$$= 370R_C - 23{,}200.$$

$$\int_0^{20} M \frac{\partial M}{\partial R_C} dx = \int_0^{20} (41 \cdot 4x - 0 \cdot 43xR_C - x^2)(-0 \cdot 43x) dx$$
$$= 490R_C - 30{,}300.$$

36

Summing and equating to zero

$$\int_0^l M\frac{\partial M}{\partial R_C}dx = 860R_C - 53,500 = 0. \qquad \therefore \ R_C = +62\cdot2 \ tons.$$

As has been shown in Chapter 2, Problems **3.1** and **3.2** are more easily solved by the "three-moment" equation, but having mastered the procedure in the simpler problems, the student will find that the remainder of the problems in this chapter are conveniently solved by the Strain Energy method.

3.3 The frame *ABCD* has rigid joints at the corners *B* and *C*, and is pin-jointed at the supports *A* and *D*. This type of frame is now used considerably in bridge construction in steel and reinforced concrete, and is usually known as a Portal Frame.

When the frame is loaded as shown in Fig. **3.3**, the points *A* and *D* have a tendency to move apart, and the horizontal force *H* is called into play. The vertical reactions V_A and V_D can be evaluated by the methods of statics and *H* is the statically indeterminate force. The frame is indeterminate to the first degree.

The writing of the bending moment equations for this type of frame sometimes presents difficulties to the student. A sheet of paper should be used to cover the frame except the portion to the right or left of the section being considered. The moments (about the edge of the paper) *of all the forces which can be seen* must then be written down, the values of any moments applied to the structure being also included.

$$V_A = V_D = 4\cdot5 \ tons. \qquad EI \ \text{is constant.}$$

Member	Bending Moment (*M*)	$\partial M/\partial H$	Limits
AB	$-Hy$	$-y$	0 to 15
CD	$-Hy$	$-y$	0 to 15
BC (left)	$-15H+4\cdot5x$	-15	0 to 10
BC (right)	$-15H+4\cdot5x$	-15	0 to 10

$$\frac{2}{EI}\int_0^{15} M\frac{\partial M}{\partial H}dy = \frac{2}{EI}\int_0^{15} Hy^2 dy = \frac{2250}{EI}H$$

$$\frac{2}{EI}\int_0^{10} M\frac{\partial M}{\partial H}dx = \frac{2}{EI}\int_0^{10}(4\cdot5x - 15H)(-15)dx = \frac{4500}{EI}H - \frac{6750}{EI}.$$

Summing and equating to zero,

$$\int_0^l M\frac{\partial M}{\partial H}dx = 6750H - 6750. \qquad H=1 \ ton \ \text{in the direction assumed.}$$

3.3

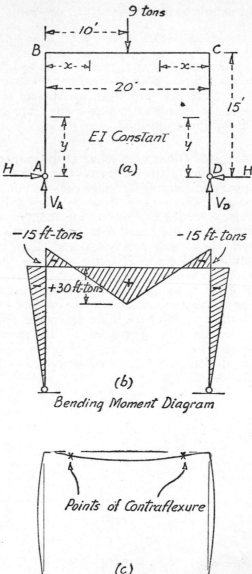

9 tons

10'

B   C

x   x

20'

15'

EI Constant

y   y

H   A   *(a)*   D   H

V_A V_D

−15 ft-tons   −15 ft-tons

+30 ft-tons +

(b)

Bending Moment Diagram

Points of Contraflexure

(c)

Deflected Form

3.3

38

Since the joints B and C are considered to be rigid, the bending moments in the beam and column at these points must be equal. The value of the bending moment at B and C is (from the table) $-15H=-15$ *ft.-tons*.

It is usual to plot the bending moment diagram on that side of the beam or column which is in tension.

The student should make a practice of drawing both the bending moment diagram and a diagram showing, to an exaggerated degree, the deflected form of the truss or frame. Points of contraflexure occur where the bending moment diagram crosses the outline of the frame.

The work done by direct stress has not been included in the total strain energy, but the stress at any section is the sum of that due to the bending moment (**b**) and that due to the direct stress, which must not be forgotten when stresses are being evaluated. At B, for example, the bending stress is obtained from the flexure formula, and the direct stress in the column AB is V_A divided by the cross-sectional area of AB.

Exercises

Solve, by *Strain Energy*, problems 2.1, 2.2, 2.3, 2.5, 2.6, 2.7, 2.8, 2.9, 2.10.

3.4 In this portal frame the moment of inertia of the beam is twice that of the columns, whose moments of inertia are equal. It is assumed that the supports yield under load, so that A and D move horizontally, relative to each other, increasing the distance AD by 0·5 in.

In such a case $\partial \overline{U}/\partial H$ represents the relative movement of the points of application of H in the direction of the line of action of H. Also, EI does not cancel out, but must be evaluated.

Moment of inertia of the columns $= 55\cdot6$ in.^{4} $(=I)$.

Moment of inertia of the beam $= 111\cdot2$ in.^{4} $(=2I)$.

Modulus of elasticity $= 13,500$ tons per sq. in. (E).

$EI = 13,500 \times 12^2 \times 55\cdot6/12^4 = 5220$ *tons-ft.*^{2}

Taking moments about D, where bending moment is zero,

$$12V_A - 12 \times 9 = 0. \quad V_A = 9 \ tons. \quad V_D = 3 \ tons.$$

Member	Bending Moment (M)	$\partial M/\partial H$	Limits	Moment of Inertia
AB	$-Hy$	$-y$	0 to 15	I
CD	$-Hy$	$-y$	0 to 15	I
BC (left)	$9x - 15H - x^2$	-15	0 to 6	$2I$
BC (right)	$3x - 15H$	-15	0 to 6	$2I$

AB and CD: $\quad \dfrac{2}{EI}\displaystyle\int_0^{15} M\dfrac{\partial M}{\partial H}dy = \dfrac{2}{EI}\int_0^{15} Hy^2 dy = \dfrac{1}{EI}(2250H).$

BC (left): $\quad \dfrac{1}{2EI}\displaystyle\int_0^{6} M\dfrac{\partial M}{\partial H}dx = \dfrac{1}{2EI}\int_0^{6}(9x - 15H - x^2)(-15)dx$

$$= \dfrac{1}{EI}(675H - 675).$$

BC (right): $\quad \dfrac{1}{2EI}\displaystyle\int_0^{6} M\dfrac{\partial M}{\partial H}dx = \dfrac{1}{2EI}\int_0^{6}(3x - 15H)(-15)dx$

$$= \dfrac{1}{EI}(675H - 405).$$

Summing the three values, and equating to the movement of A and D, relative to each other, in a horizontal direction,

$$\dfrac{1}{5220}(3600H - 1080) = -1/24 \quad \therefore \ H = 0\cdot24 \ tons,$$

whence, by substitution in the "bending moment" column of the table, the bending moment diagram can be drawn.

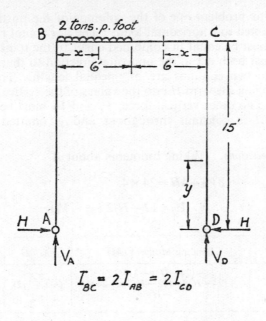

$$I_{BC} = 2I_{AB} = 2I_{CD}$$

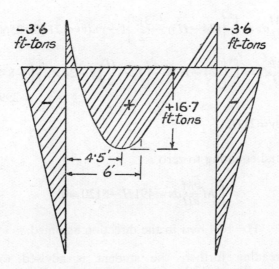

Bending Moment Diagram

3.4

3.5 In this problem one of the columns of the portal frame is subjected to a horizontal load. The horizontal reactions at the hinges must be equal in combined effect to the total horizontal load and must both act in the opposite direction to that load.

Again, the two columns are of unequal length. This has the effect of bringing the term H into the values of the vertical reactions. Since there is no other vertical force, V_L and V_R must be equal and opposite. EI is constant throughout and is omitted from the equation.

Vertical Reactions. Taking moments about A

$$8V_R + 4H = 24 \times 4.$$

$$V_R = 12 - H/2 \ (= -V_L).$$

Member	Bending Moment (M)	$\partial M/\partial H$	Limits
AB	$+(24-H)y - 3y^2/2$	$-y$	0 to 8
BC	$+(24-H)8 - (12-H/2)x - 24 \times 4$	$(-8+x/2)$	0 to 8
CD	$-Hy$	$-y$	0 to 4

$$\int_0^8 M \frac{\partial M}{\partial H} dy = \int_0^8 \left\{ (24-H)y - \frac{3y^2}{2} \right\} (-y) dy = 171H - 2560.$$

$$\int_0^8 M \frac{\partial M}{\partial H} dx = \int_0^8 \left\{ (24-H)8 - \left(12 - \frac{H}{2}\right)x - 24 \times 4 \right\} \left(-8 + \frac{x}{2}\right) dx$$
$$= 299H - 2560.$$

$$\int_0^4 Hy^2 dy = 21H.$$

Summing and equating to zero

$$\int_0^l M \frac{\partial M}{\partial H} ds = 491H - 5120 = 0.$$

$H = 10 \cdot 4$ *tons* in the direction assumed.

Before reading further, the student is advised to become thoroughly familiar with the solution of frames which are statically indeterminate to the first degree. Problems **3.1** to **3.5** should be solved without reference to the text, and for each frame the bending moment diagram and deflected form should be drawn.

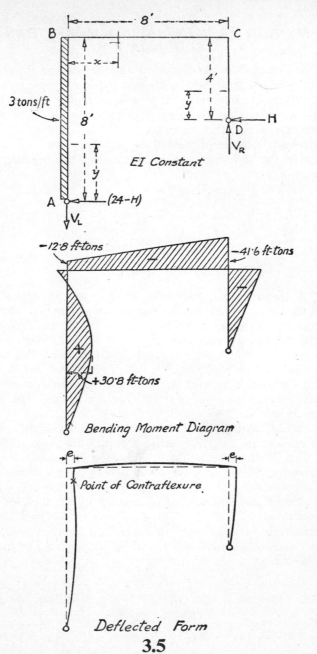

$3\,tons/ft$

EI Constant

$(24-H)$

$-12·8\,ft\text{-}tons$ $-41·6\,ft\text{-}tons$

$+30·8\,ft\text{-}tons$

Bending Moment Diagram

Point of Contraflexure

Deflected Form

3.5

3.6

BEAMS AND FRAMES HAVING MORE THAN ONE REDUNDANCY

So far, this chapter has discussed only beams and frames having one statically indeterminate reaction. If the structure to be solved is statically indeterminate to the second or higher degree, then two or more equations are required in order to evaluate the statically indeterminate forces and moments.

These equations are obtained by studying the deflection and rotation of the supports as was indicated at the beginning of the chapter.

3.6 A beam fixed horizontally at each end is statically indeterminate to the second degree, for either two moments or a force and a moment must be released before the beam is in a statically determinate condition.

The beam is of steel with a moment of inertia of 55·6 in.^{4}, as in Problem 3·4. $EI = 5220$ *tons-ft.*^{2}

When the load is applied the left-hand support rotates towards the beam through an angle of 0·0383 radians without sinking below its original level.

Assume M_L and R_L as the statically indeterminate quantities. If it is not certain in which sense a supporting force or moment acts, a direction may be assumed, and if the magnitude of the force or moment is found to have a positive sign, then the assumed direction was the correct one. In this instance the directions shown in the figure will be used in drawing up the table of bending moments.

Alternatively (as was done in Problem **2.2**) M_L and R_L might both be assumed positive, when the signs appearing in the results would be correct.

Member	Bending Moment (M)	$\partial M/\partial R_L$	$\partial M/\partial M_L$	Limits
AB	$-M_L + R_L x - x^2$	$+x$	-1	0 to 15
BC	$-M_L + R_L x - 30(x-7\cdot5)$	$+x$	-1	15 to 20

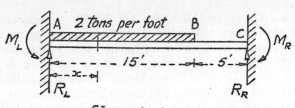

EI Constant

3.6

First Equation.

$$\frac{1}{EI}\int_0^{20} M\frac{\partial M}{\partial M_L}dx = -0.0383.$$

$$\frac{1}{EI}\int_0^{15} M\frac{\partial M}{\partial M_L}dx = \frac{1}{EI}\int_0^{15}(-M_L+R_Lx-x^2)(-1)dx$$

$$=\frac{1}{EI}(15M_L-113R_L+1125).$$

$$\frac{1}{EI}\int_{15}^{20} M\frac{\partial M}{\partial M_L}dx = \frac{1}{EI}\int_{15}^{20}\{-M_L+R_Lx-30(x-7.5)\}(-1)dx$$

$$=\frac{1}{EI}(5M_L-87R_L+1500).$$

Summing: $+20M_L-200R_L+2625 = -5220\times0.0383.$

Second Equation.

$$\frac{1}{EI}\int_0^{20} M\frac{\partial M}{\partial R_L}dx = 0.$$

$$\frac{1}{EI}\int_0^{15} M\frac{\partial M}{\partial R_L}dx = \frac{1}{EI}\int_0^{15}(-M_L+R_Lx-x^2)xdx$$

$$=\frac{1}{EI}(-113M_L+1125R_L-12{,}656).$$

$$\frac{1}{EI}\int_{15}^{20} M\frac{\partial M}{\partial R_L}dx = \frac{1}{EI}\int_{15}^{20}\{M_L+R_Lx-30(x-7.5)\}xdx$$

$$=\frac{1}{EI}(-87M_L+1542R_L-26{,}563).$$

Summing: $-200M_L+2667R_L-39{,}219=0.$

Simultaneous solution of the two equations gives

$$R_L=16.5 \text{ tons.} \quad M_L=23.7 \text{ ft.-tons.} \quad M_R=69 \text{ ft.-tons.}$$

The positive signs indicate that the assumed directions for M_L and R_L are the correct ones. Draw the bending moment and shear force diagrams.

3.7 The columns of this frame are fixed in a vertical direction at the bottom. If one of these fixed ends is released so that the frame becomes statically determinate, M_D, V_D and H cease to act, and the frame is thus indeterminate to the third degree and requires three equations for its solution. EI is constant throughout.

In order that only three unknown quantities should occur in these equations it is most convenient to determine all the bending moment equations from one end of the frame.

Member	Bending Moment (M)	$\partial M/\partial H$	$\partial M/\partial V_D$	$\partial M/\partial M_D$	Limits
DC	$M_D - Hy$	$-y$	—	$+1$	0 to 6
CB (right)	$M_D - 6H + V_D x$	-6	$+x$	$+1$	0 to 6
CB (left)	$M_D - 6H + V_D x - 4(x-6)$	-6	$+x$	$+1$	6 to 8
BA	$M_D - Hy + 8V_D - 8$	$-y$	$+8$	$+1$	0 to 6

First Equation.

$$\frac{1}{EI}\int_0^l M\frac{\partial M}{\partial H}ds=0.$$

$$\int_0^6 (Hy^2 - M_D y)dy = 72H - 18M_D.$$

$$\int_0^6 (36H - 6M_D - 6V_D x)dx = 216H - 36M_D - 108V_D.$$

$$\int_6^8 (36H - 6M_D - 6V_D x + 24x - 144)dx = 72H - 12M_D - 84V_D + 48.$$

$$\int_0^6 (Hy^2 - M_D y - 8V_D y + 8y)dy = 72H - 18M_D - 144V_D + 144.$$

Summing and equating to zero,
$$+432H - 84M_D - 336V_D + 192 = 0.$$

Second Equation.

$$\frac{1}{EI}\int_0^l M\frac{\partial M}{\partial V_D}ds=0.$$

$$\int_0^6 (M_D x - 6Hx + V_D x^2)dx = -108H + 18M_D + 72V_D.$$

$$\int_6^8 \{M_D x - 6Hx + V_D x^2 - 4x(x-6)\}dx = -84H + 14M_D + 99V_D - 59.$$

$$\int_0^6 (8M_D - 8Hy + 64V_D - 64)dy = -144H + 48M_D + 384V_D - 384.$$

Summing and equating to zero,
$$-336H + 80M_D + 555V_D - 443 = 0.$$

46

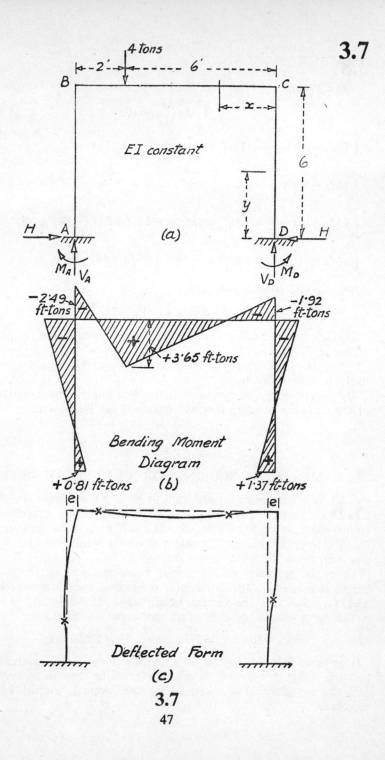

4 tons

B ←—— 2' ——→ ←———— 6' ————→ C

←— x —→

EI constant

6

y

H A (a) D H

M_A V_A V_D M_D

−2·49 ft-tons   −1·92 ft-tons

−

−

+

+3·65 ft-tons

Bending Moment
Diagram

+0·81 ft-tons   (b)   +1·37 ft-tons

|e| |e|

Deflected Form
(c)

3.8

Third Equation.

$$\frac{1}{EI}\int_0^{l} M\frac{\partial M}{\partial M_D}ds=0.$$

$$\int_0^6 (M_D-Hy)dy=-18H+6M_D.$$

$$\int_0^6 (M_D-6H+V_Dx)dx=-36H+6M_D+18V_D.$$

$$\int_6^8 \{M_D-6H+V_Dx-4(x-6)\}dx=-12H+2M_D+14V_D-8.$$

$$\int_0^6 (M_D-Hy+8V_D-8)dy=-18H+6M_D+48V_D-48.$$

Summing and equating to zero,

$$-84H+20M_D+80V_D-56=0.$$

Solving these equations simultaneously, the following results are obtained:

$$H=0{\cdot}55 \ tons; \quad V_D=0{\cdot}93 \ tons; \quad M_D=1{\cdot}37 \ ft.\text{-}tons$$

in the assumed directions.

By substituting these values in the bending moment equations of the table the bending moment diagram can be drawn.

FRAMES WITH SLOPING OR CURVED MEMBERS

3.8 This frame has a rigid joint at B, and the members AB and BC are inclined to the horizontal. The procedure is similar to that of Problem 3.3. It must always be remembered that the integrations must be taken *along the members of the frame, and must cover the whole of the frame.*

From the geometry of the frame, a distance s along the sloping length is related to the horizontal projection x by the expression $x=0{\cdot}8s$. Similarly the vertical height $y=0{\cdot}75x=0{\cdot}6s$.

Taking moments about C of all the forces on the frame,

$$8V_A=30. \quad V_A=3{\cdot}75 \ tons. \quad V_C=1{\cdot}25 \ tons.$$

The table is made up as in previous problems, and then the bending moment expressions are converted to read in terms of s, and the integration is taken along the sloping lengths of the members.

Member	Bending Moment	Bending Moment re-written	$\partial M/\partial H$	Limits
AB (left)	$3{\cdot}75x-Hy$	$3s-0{\cdot}6sH$	$-0{\cdot}6s$	0 to 2·5
AB (right)	$3{\cdot}75x-Hy-5(x-2)$	$3s-0{\cdot}6sH-4s+10$	$-0{\cdot}6s$	2·5 to 5
BC	$1{\cdot}25x-Hy$	$s-0{\cdot}6sH$	$-0{\cdot}6s$	0 to 5

Only one equation is required:

$$30H=68{\cdot}6. \quad H=+2{\cdot}29 \text{ tons in the assumed direction.}$$

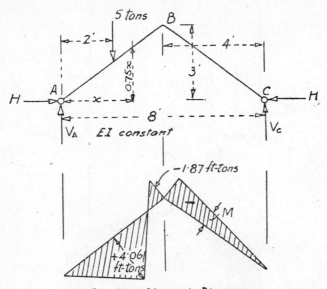

Bending Moment Diagram

3.8

3.9 This rigid building frame carries a 10-ton load on a bracket projecting from one of the columns. The frame is symmetrical about a vertical centre line, and the moments of inertia of the various lengths are in the ratios shown in (a).

Taking moments about E of all forces and moments on the frame,

$$16V_A=10\times15. \quad V_A=9{\cdot}4 \text{ tons.} \quad V_E=0{\cdot}6 \text{ tons.}$$

In this instance, the sloping lengths BC and CD each $=\sqrt{80}$ ft. Thus $x=0{\cdot}894s$, and $z=0{\cdot}447s$.

3.9

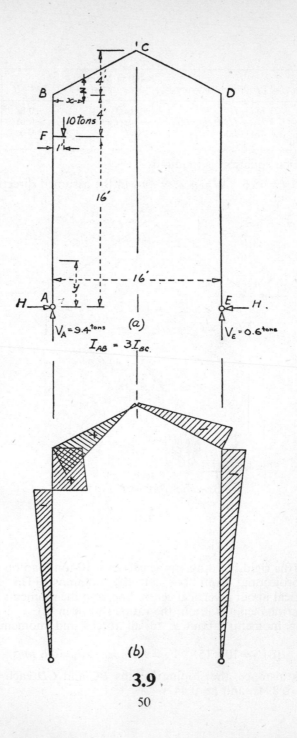

C

B
4'
3
x-b
4'
D
10 tons
F
1'
16'
16'
H — A
y
E — H.
$V_A = 9.4^{tons}$ (a) $V_E = 0.6^{tons}$

$I_{AB} = 3I_{BC}$

(b)

The table is made up in the usual way:

Member	Bending Moment	Bending Moment re-written	Moment of Inertia	$\partial M/\partial H$	Limits
AF	$-Hy$	$-Hy$	$3I$	$-y$	0–16
FB	$-Hy+10$	$-Hy+10$	$3I$	$-y$	16–20
BC	$-H(20+z)+9{\cdot}4x$ $-10(x-1)$	$-H(20+0{\cdot}447s)$ $+9{\cdot}4(0{\cdot}894s)$ $-10(0{\cdot}894s-1)$	I	$-(20$ $+0{\cdot}447s)$	0–$\sqrt{80}$
DE	$-Hy$	$-Hy$	$3I$	$-y$	0–20
CD	$-H(20+z)+0{\cdot}6x$	$-H(20+0{\cdot}447s)$ $+0{\cdot}6(0{\cdot}894s)$	I	$-(20$ $+0{\cdot}447s)$	0–$\sqrt{80}$

The equation from which the value of H may be determined is $\int \dfrac{M}{EI}\dfrac{\partial M}{\partial H}ds=0$, since there is no relative horizontal movement of the abutments. It is to be noted that integrations should be taken along the centre lines of the members. Although various "short cuts" and approximations may be made, it is better for the beginner to carry out the integration along the sloping lengths in order to emphasise this part of the principle on which the solution of these problems depends.

$$AF \quad \frac{1}{3I}\int_0^{16} Hy^2 dy = (455H)\frac{1}{I}.$$

$$FB \quad \frac{1}{3I}\int_{16}^{20}(Hy^2-10y)dy = (434H-240)\frac{1}{I}.$$

$$BC \quad \frac{1}{I}\int_0^{\sqrt{80}}\{H(20+{\cdot}447s)-9{\cdot}4({\cdot}894s)+10({\cdot}894s-1)\}(20+{\cdot}447s)ds$$

$$= (4341H-1590)\frac{1}{I}.$$

$$DE \quad \frac{1}{3I}\int_0^{20} Hy^2 dy = (889H)\frac{1}{I}.$$

$$CD \quad \frac{1}{I}\int_0^{\sqrt{80}}\{H(20+{\cdot}447s)-0{\cdot}6({\cdot}894s)\}(20+{\cdot}447s)ds$$

$$= (4341H-491)\frac{1}{I}.$$

$$10,460H = 2221.$$

Summing and equating to zero, $H=0{\cdot}21$ *tons*.

Evaluate the various critical ordinates of the bending moment diagram shown in (b).

3.10 In solving curved statically indeterminate frames it is important to remember that the integration must be carried out over the whole length of the frame. The elemental length ds then becomes $r d\theta$ in the case of a circular frame.

Consider the semicircular arch shown in Fig. **3.10**. It is statically indeterminate to the first degree. $V_A = V_C = 1$ *ton*.

Member	Bending Moment	$\partial M/\partial H$	Limits
CB	$(6-6\cos\theta)V_C - 6\sin\theta H$	$-6\sin\theta$	0 to $\pi/2$
AB	$(6-6\cos\theta)V_A - 6\sin\theta H$	$-6\sin\theta$	0 to $\pi/2$

$$12\int_0^{\pi/2}\{6(1-\cos\theta)-6H\sin\theta\}(-6\sin\theta)d\theta=0.$$

$$432\int_0^{\pi/2}(-\sin\theta+\cos\theta\sin\theta+H\sin^2\theta)d\theta$$

$$=432\left[\cos\theta+\frac{\sin^2\theta}{2}+H\frac{\theta}{2}-H\frac{\sin 2\theta}{4}\right]_0^{\pi/2}=0.$$

$$H=2/\pi \text{ tons.}$$

By substituting this value of H in the bending moment equations of the table, the bending moment for any point on the arch ring can be determined.

Arches, such as that considered in this problem, are more generally solved by the variation of the Method of Strain Energy which is described in Chapter 4.

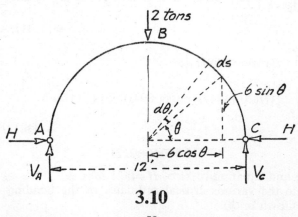

3.10

52

MULTI-SPAN, MULTI-STOREY, VIERENDEEL AND CIRCULAR FRAMES

3.11 If three of the supporting reactions of this frame are know it is statically determinate. Three equations are therefore required beyond those supplied by the methods of statics.

The six values of H and V must be reduced to three before the usual table can be drawn up. H_1 and H_3 evidently act inward to prevent the columns splaying under the load. The direction of H_2 may be less evident, but this is unimportant. The value of H_2 must be such that the horizontal forces balance each other, or

$$H_2 = H_1 - H_3.$$

For vertical equilibrium, the sum of the vertical reactions must equal the vertical loads, or $V_1 + V_2 + V_3 = 30$ *tons.*

Again, the moments of all the forces on the structure, taken about any point, must be zero. Take moments about the hinge F since this eliminates H_3 and V_3.

$$20V_1 + 10V_2 = 20 \times 15 + 10 \times 5.$$

$$2V_1 + V_2 = 35 \text{ } tons.$$

Eliminating V_2 by substituting $V_2 = 30 - V_1 - V_3$

$$V_1 = V_3 + 5.$$

The table can now be drawn up so that only H_1, H_3 and V_3 appear as unknown terms.

Member	Bending Moment	Moment re-written	Moment of Inertia	$\frac{\partial M}{\partial H_1}$	$\frac{\partial M}{\partial H_3}$	$\frac{\partial M}{\partial V_3}$	Limits
AB	$-H_1 y$	$-H_1 y$	I	$-y$	...	...	0–6
BC	$-6H_1 + V_1 x - x^2$	$-6H_1 + (V_3+5)x$ $-x^2$	$2I$	-6	...	$+x$	0–10
CD	$-H_2 y$	$(-H_1 + H_3)y$	I	$-y$	$+y$	...	0–6
EC	$-6H_3 + V_3 x - x^2/2$	$-6H_3 + V_3 x$ $-x^2/2$	$2I$	...	-6	$+x$	0–10
EF	$-H_3 y$	$-H_3 y$	I	...	$-y$	...	0–6

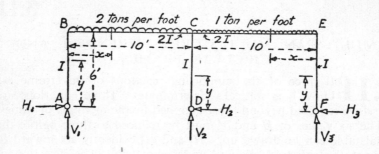

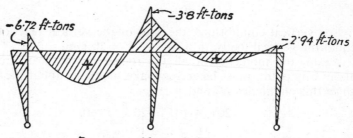

Bending Moment Diagram

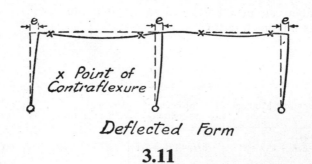

x Point of
Contraflexure

Deflected Form

3.11

First Equation.

$$\frac{1}{EI}\int_0^l M\frac{\partial M}{\partial H_1}ds=0.$$

$$\frac{1}{EI}\int_0^6 H_1 y^2 dy=(72H_1)\frac{1}{EI}.$$

$$\frac{1}{2EI}\int_0^{10}(36H_1-6V_3x-30x+6x^2)dx=(180H_1-150V_3+250)\frac{1}{EI}.$$

$$\frac{1}{EI}\int_0^6 (H_1-H_3)y^2 dy=(72H_1-72H_3)\frac{1}{EI}.$$

Summing and equating to zero,

$$+324H_1-72H_3-150V_3+250=0.$$

Second Equation.

$$\frac{1}{EI}\int_0^l M\frac{\partial M}{\partial H_3}ds=0.$$

$$\frac{1}{EI}\int_0^6 (-H_1+H_3)y^2 dy=(-72H_1+72H_3)\frac{1}{EI}.$$

$$\frac{1}{2EI}\int_0^{10}(36H_3-6V_3x+3x^2)dx=(180H_3-150V_3+500)\frac{1}{EI}.$$

$$\frac{1}{EI}\int_0^6 H_3 y^2 dy=(72H_3)\frac{1}{EI}.$$

Summing and equating to zero,

$$-72H_1+324H_3-150V_3+500=0.$$

Third Equation.

$$\frac{1}{EI}\int_0^l M\frac{\partial M}{\partial V_3}ds=0.$$

$$\frac{1}{2EI}\int_0^{10}(-6H_1x+V_3x^2+5x^2-x^3)dx=(-150H_1+167V_3-417)\frac{1}{EI}.$$

$$\frac{1}{2EI}\int_0^{10}\left(-6H_3x+V_3x^2-\frac{x^3}{2}\right)dx=(-150H_3+167V_3-625)\frac{1}{EI}.$$

Summing and equating to zero,

$$-150H_1-150H_3+333V_3-1042=0.$$

3.11

By a simultaneous solution of the first, second and third equations the values of the statically indeterminate quantities, H_1, H_3 and V_3 can be found:

$$H_1 = 1 \cdot 12 \; tons. \quad H_3 = 0 \cdot 49 \; tons. \quad V_3 = 3 \cdot 86 \; tons.$$

Hence $H_2 = H_1 - H_3 = 1 \cdot 12 - 0 \cdot 49 = 0 \cdot 63$ *tons* in the direction assumed (towards the left).

$$V_1 = V_3 + 5 = 8 \cdot 86 \; tons. \quad V_2 = 30 - 3 \cdot 86 - 8 \cdot 86 = 17 \cdot 28 \; tons.$$

By taking moments of all the forces on one side of critical sections the bending moment diagram may be drawn. The student is advised to make this calculation for himself.

CONVENTION OF SIGNS

The convention of signs used in Chapters 2 and 3 is convenient so long as the structure under consideration is a simple beam or portal frame. When, however, double portals (as in the present example) or more complex frames require solution, this convention of signs becomes clumsy. For instance, it is obvious that the sum of the moments at C should be zero if the joint is to remain in equilibrium. Although, numerically, the moment in CB balances the sum of the moments in CE and CD yet the signs do not help when this check is made. Similarly, moments at corners, such as B and E, should have opposite signs if the sum is to be zero. This latter error, however, is not important, and the convention of Chapters 2 and 3 may be used for simple portals without inconvenience.

When more complex frames are considered a more useful convention must be found. As an introduction to future work, the student is recommended to read the first few paragraphs of Chapter 5, and then study Problems **3.12** and **3.13**. In these problems the calculation will be carried out according to the convention of Chapter 2, but the bending moment diagram will be marked according to the convention of Chapter 5. A careful comparison of the two methods of determining the signs should be made.

Exercises

Solve, by *Strain Energy*, problems 5.1, 5.5, 6.1, 6.2, 6.3.

3.12 When frames (such as building frames) which possess numerous indeterminate forces and moments are considered, it is more convenient to choose moments rather than horizontal and vertical forces as the indeterminate quantities. This problem shows how the values of statically indeterminate moments may be found by the Method of Strain Energy.

At first sight (a) it appears that twelve moments must be determined. However, since the loading is symmetrical about the vertical axis of the frame, and the construction of the frame, both in linear and cross-sectional dimensions, is symmetrical, only one side of the frame need be considered. This reduces the number of unknown moments to six.

Further, since each joint of the structure is in equilibrium under the action of the moments shown in (a), the following statements are true (using the sign convention of Chapters 2 and 3).

1. $M_5 = M_6$.   2. $M_4 = M_2 + M_3$.   3. $M_1 = -M_2/2$.

This finally reduces the number of unknown moments to three, the three chosen for solution being M_2, M_3 and M_5.

Equation 3, above, could not be used if the loading or the structure of the frame were unsymmetrical about the centre line. The statement (3) may be put into words by saying that when a moment is applied at the free end of a member fixed at the other end, *and there is no relative movement of the ends*, then the moment at the fixed end is half of that at the free end and in the same direction (clockwise or counter-clockwise). With the convention of Chapters 2 and 3 these moments will be of opposite sign, and with the convention of Chapter 5, of the same sign. If the frame were not symmetrical or the loading were eccentric, then the condition of no relative movement between the ends would not be fulfilled. The frame would sway laterally and the relationship between the moments at, say, A and B would not be known.

If a member is acted on by a bending moment at both ends, the bending moment at any section, distant x from the end X, is

$$M_x - (M_x - M_y)x/l.$$

This can be verified by reference to the three possible conditions shown in (d). If, in addition, the member carries a load, the "free" bending moment due to this load must be added.

3.12

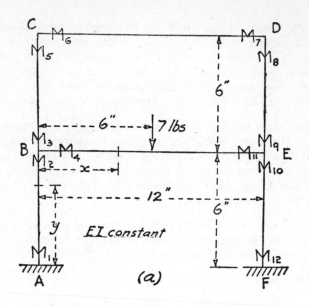

(a)

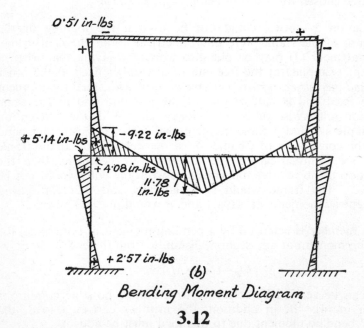

(b)

Bending Moment Diagram

3.12

The table is constructed by writing down the bending moment at any point in each member in terms of the unknown moments.

Member	Bending Moment	Bending Moment re-written	$\frac{\partial M}{\partial M_2}$	$\frac{\partial M}{\partial M_3}$	$\frac{\partial M}{\partial M_5}$
AB	$M_1-(M_1-M_2)y/6$	$M_2\left(-\frac{1}{2}+\frac{y}{4}\right)$	$\left(-\frac{1}{2}+\frac{y}{4}\right)$	—	—
BC	$M_3-(M_3-M_5)y/6$	$M_3\left(1-\frac{y}{6}\right)+M_5\frac{y}{6}$	—	$\left(1-\frac{y}{6}\right)$	$\frac{y}{6}$
CD	$M_6-(M_6-M_7)x/12$	M_5	—	—	$+1$
BE	$M_4-(M_4-M_{11})x/12+3\cdot5x$	$M_2+M_3+3\cdot5x$	$+1$	$+1$	—

In obtaining the following three equations, integration is taken over half the span only:

First Equation.

$$\frac{1}{EI}\int M\frac{\partial M}{\partial M_2}ds=0.$$

$$\int_0^6 M_2\left(-\frac{1}{2}+\frac{y}{4}\right)^2 dy=1\cdot5M_2.$$

$$\int_0 (M_2+M_3+3\cdot5x)dx=6M_2+6M_3+63.$$

Summing and equating to zero,

$$7\cdot5M_2+6M_3+63=0.$$

Second Equation.

$$\frac{1}{EI}\int M\frac{\partial M}{\partial M_3}ds=0.$$

$$\int_0^6\left\{M_3\left(1-\frac{y}{6}\right)^2+M_5\frac{y}{6}\left(1-\frac{y}{6}\right)\right\}dy=2M_3+M_5.$$

$$\int_0^5 (M_2+M_3+3\cdot5x)dx=6M_2+6M_3+63.$$

Summing and equating to zero,

$$6M_2+8M_3+M_5+63=0.$$

3.12

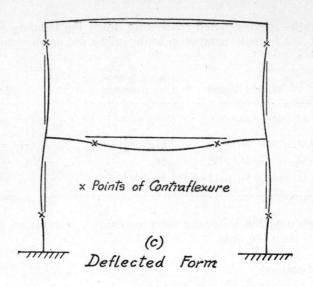

× Points of Contraflexure

(c)
Deflected Form

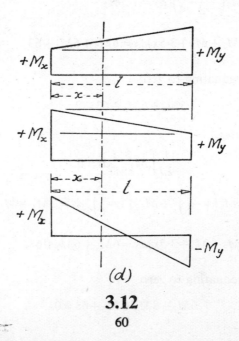

$+M_x$ $+M_y$

l

x

$+M_x$ $+M_y$

x l

$+M_x$ $-M_y$

(d)

3.12

60

Third Equation.

$$\frac{1}{EI}\int M\frac{\partial M}{\partial M_5}ds=0.$$

$$\int_0^6\left\{M_3\left(1-\frac{y}{6}\right)\frac{y}{6}+M_5\frac{y^2}{36}\right\}dy=M_3+2M_5.$$

$$\int_0^6 M_5dx=6M_5.$$

Summing and equating to zero,

$$M_3+8M_5=0.$$

By a simultaneous solution of these three equations, the statically indeterminate moments M_2, M_3 and M_5 are obtained, and the bending moment diagram for the frame drawn.

$M_3=-4\cdot08$ *in.-lb.* $M_5=+0\cdot51$ *in.-lb.* $M_2=-5\cdot14$ *in.-lb.*

From the known relationships between the moments,

$$M_4=M_2+M_3=-9\cdot22 \text{ } in.\text{-}lb.$$

$$M_1=-\frac{M_2}{2}=+2\cdot57 \text{ } in.\text{-}lb.$$

It is to be noted that the signs used in the bending moment diagram **(b)** are according to the convention of Chapter 5. The two conventions should be carefully compared, in order that the work in subsequent chapters may be well understood.

Exercises

Solve, by *Strain Energy*, problems 6.4, 7.1, 7.2, 6.5, 7.3, 5.4, 5.6, 5.7, 5.8, 6.6, 6.7, 6.8, 6.9, 6.11, 6.12, 7.4, 7.6.

3.13 *Vierendeel Frame.* A rectangular frame of this type is frequently used in the construction of box culverts, and is usually loaded symmetrically by earth pressure and loading on both sides and top. In order to illustrate the method of solution by means of Strain Energy analysis, a single concentrated load has been applied at the centre of the upper beam.

As in Problem **3.12**, it is convenient to consider the bending moments at the corners of the frame as the statically indeterminate factors. Of the eight unknown moments only two need be found directly. The others can be evaluated from the symmetry of the frame and from the following two equalities:

$$M_3 = M_4. \quad M_1 = M_2.$$

The table is made up in a similar manner to that of Problem **3.12**.

Member	Bending Moment	Bending Moment re-written	$\dfrac{\partial M}{\partial M_2}$	$\dfrac{\partial M}{\partial M_3}$	Limits	Moment of Inertia
AB	$M_2-(M_2-M_3)\dfrac{y}{6}$	$+M_2\left(1-\dfrac{y}{6}\right)+M_3\dfrac{y}{6}$	$\left(1-\dfrac{y}{6}\right)$	$\dfrac{y}{6}$	0–6	$2I$
BC	$M_4-(M_4-M_5)\dfrac{x}{8}$ $+2x$	$+M_3+2x$	—	$+1$	0–4	I
AD	$M_1-(M_1-M_8)\dfrac{x}{8}$	$+M_2$	$+1$	—	0–4	$4I$

First Equation.
$$\frac{1}{EI}\int M\frac{\partial M}{\partial M_2}ds=0.$$

$$\frac{1}{2I}\int_0^6\left\{M_2\left(1-\frac{y}{6}\right)^2+M_3\frac{y}{6}\left(1-\frac{y}{6}\right)\right\}dy=(M_2+0\cdot5M_3)\frac{1}{I}.$$

$$\frac{1}{4I}\int_0^4 M_2dx=\frac{M_2}{I}.$$

Summing and equating to zero,
$$2M_2+0\cdot5M_3=0.$$

Second Equation.
$$\frac{1}{EI}\int M\frac{\partial M}{\partial M_3}ds=0.$$

$$\frac{1}{2I}\int_0^6\left(M_2\frac{y}{6}-M_2\frac{y^2}{36}+M_3\frac{y^2}{36}\right)dy=(0\cdot5M_2+M_3)\frac{1}{I}.$$

$$\frac{1}{I}\int_0^4(M_3+2x)dx=(4M_3+16)\frac{1}{I}.$$

Summing and equating to zero,
$0\cdot5M_2+5M_3+16=0$, whence $M_2=+0\cdot82$ *ft.-tons* and $M_3=-3\cdot28$ *ft.-tons.*

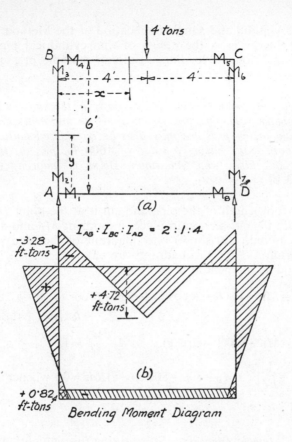

4 tons

$I_{AB} : I_{BC} : I_{AD} = 2 : 1 : 4$

−3·28 ft-tons

+4·72 ft-tons

+0·82 ft-tons

(a)

(b)

Bending Moment Diagram

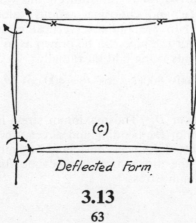

(c)

Deflected Form.

$\mathbf{3.14}$ A useful and simple application of the Method of Strain Energy is in the design of thin cylindrical pipes under different conditions of loading, assuming that the pressure inside the pipe is atmospheric.

A reinforced concrete pipe is of uniform thickness and 24 in. mean diameter. The loading from above is uniformly distributed, and the supporting pressure may also be considered uniformly distributed, both loads having a value of 1600 lb. per sq. ft. If the pipe is 2 in. thick, what maximum stress is induced? (Refer to Problem 3.10.)

Take a foot length of the pipe and cut it at the point D. Then, to maintain the cut section in place, a moment M_D and a reaction V_D are required. (b) shows one quadrant of the pipe supported in equilibrium. All four quadrants are alike.

The bending moment at X is

$$M_X = M_D - V_D(R - R \cos \theta) + \frac{w}{2}(R - R \cos \theta)^2 \ ft.\text{-}lb. :$$

$$R = 1 \ ft. \quad V_D = 1600 \ lb.$$

$$= M_D - 800(1 - \cos^2 \theta). \qquad \frac{\partial M_X}{\partial M_D} = 1.$$

$$\frac{\partial \bar{U}}{\partial M_D} = \frac{1}{EI}\int_0^{\pi/2} (M_D - 800 - 800 \cos^2 \theta)R d\theta = 0; \quad \text{whence } M_D$$

$$= 400 \ ft.\text{-}lb.$$

Bending Moment Diagram. Substituting this value of M_D in the expression for M_X, and choosing various values of θ from 0 (horizontal radius) to $\pi/2$ radians (vertical radius), the bending moment diagram for the pipe can be drawn as in (c), the values of the bending moments being laid off radially.

$$M_X = 400 - 800(1 - \cos^2 \theta) = 400 \cos 2\theta \ ft.\text{-}lb.$$

Maximum Stress at D. The maximum stress in the concrete at D is a combination of the bending and direct stress. Although the work done by direct stress was considered negligible in the calculations for M_D, yet the direct stress itself cannot be neglected when determining maximum stress.

64

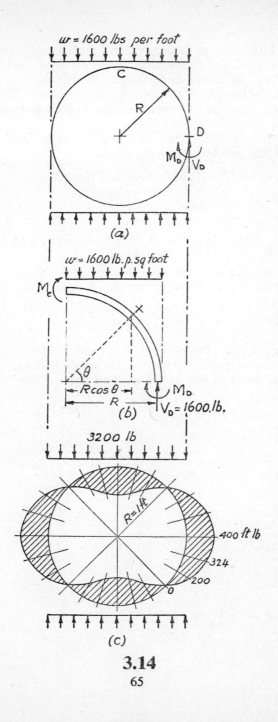

$w = 1600$ lbs per foot

C

R

M_D V_D

D

(a)

$w = 1600$ lb. p. sq foot

M_c

θ

$R\cos\theta$

R

(b)

M_D

$V_D = 1600.$ lb.

3200 lb

$R = 1$ ft

400 ft lb

324

200

0

(c)

3.14

3.15 Pipes, such as that in Problem **3.14,** may not be uniformly loaded and supported. By some error of construction the pipe may, for example, be supported only by a spur or ridge of rock at the bottom of the trench. If the pipe of Problem **3.14** were supported in this way, what increase in maximum stress would result?

Take a foot length of the pipe and cut it as in the previous problem. The bending moment at X is

$$M_X = M_D - 1600(R - R \cos \theta) \text{ ft.-lb.} \qquad R = 1 \text{ ft.}$$

$$\frac{\partial M_X}{\partial M_D} = 1.$$

$$\frac{\partial \overline{U}}{\partial M_D} = \frac{1}{EI} \int_0^{\pi/2} \{M_D - 1600(1 - \cos \theta)\} R d\theta = 0; \text{ whence } M_D = 582 \text{ ft.-lb.}$$

Bending Moment Diagram. Substituting this value of M_D in the expression for M_X the bending moment diagram can be drawn.

$$M_X = 582 - 1600(1 - \cos \theta)$$
$$= -1018 + 1600 \cos \theta.$$

Maximum Stress. The student is advised to check the value of the maximum stress at C and at D, remembering to take into account the direct as well as the bending stress.

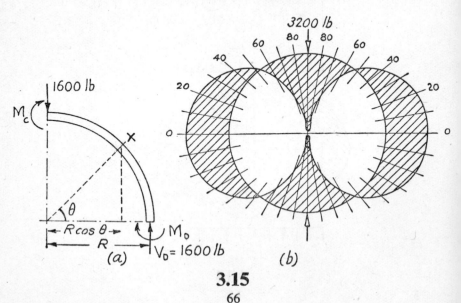

3.15

66

CHAPTER 4

TWO-HINGED AND FIXED ARCHES

Significance of the Arch Profile

A simply supported beam under normal conditions of loading is subjected to a high bending moment near the centre of the span. This requires the provision of a deep and heavy cross-section at this point, and such additional dead load increases the bending moment still further. When the beam length approaches the long span necessary for a large bridge, the additional central dead load renders this type of construction uneconomic or even impossible.

This heavy bending moment at the centre of a simply supported beam may be reduced by bending the beam into a curved shape in a vertical plane. Such a step brings into play a horizontal force H (Fig. 4.0) which, at any section of the beam, causes a bending moment of opposite sign to that caused by the superimposed loading. Fig. 4.0 shows how the bending moment at the centre of the span (and at other points along its length) is considerably reduced, thus allowing longer spans to be covered with a girder cross-section of a given moment of resistance.

Before the bending moment diagram shown in Fig. 4.0d can be drawn, the value of H must be found.

If a hinge is inserted at the crown or highest point of the arch, H can be determined by taking moments about this hinge where, from the inability of a hinge to resist rotation, the bending moment is known to be zero. Such *Three-hinged Arches* are statically determinate and are outside the scope of this book.

TWO-HINGED ARCHES

are statically indeterminate to the first degree and one equation, in addition to those of statics, must be employed in determining H. There is no fundamental difference between two-hinged arches and the two-hinged portal frames of Chapter 3.

The single equation necessary to determine H is evolved by consideration of the horizontal movement of the abutments if H were removed, and the student is referred to any standard text-book on Theory of Structures for the derivation of the expression which is

4.0

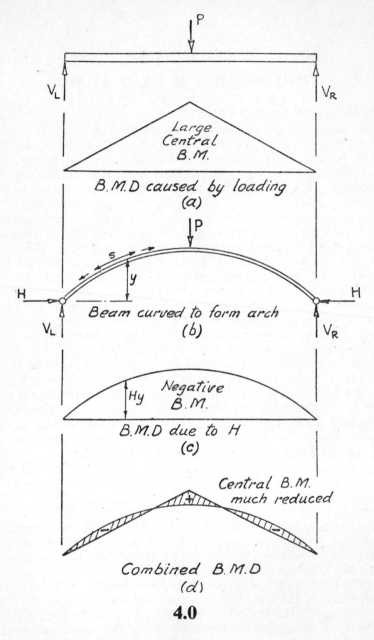

P

V_L V_R

Large
Central
B.M.

B.M.D caused by loading
(a)

P

s

y

H   Beam curved to form arch   H

V_L V_R

(b)

Hy   Negative
B.M.

B.M.D due to H
(c)

Central B.M.
much reduced

+

−   −

Combined B.M.D
(d)

4.0

applied to various problems in this chapter. The method is derived from that used in Chapter 3 and is usually more convenient than a direct application of Strain Energy procedure.

Arches may be divided into two classes:

(a) Arches of simple geometrical form, such as
    (i) segment of a circle;
    (ii) parabola.

(b) Arches whose outline is constructed of portions of several curves, or otherwise departs from a simple geometrical shape.

For the first of these types y (4.0b) can usually be evaluated as a function of s, the distance measured along the curve of the arch rib, and I is either constant or is also a function of s. In such instances the determination of the value of H can be carried out by integration over the length of the curved arch rib; and

$$H = \frac{\int My\frac{ds}{I}}{\int y^2\frac{ds}{I}}$$

For the second of these types

$$H = \frac{\sum My\frac{\delta s}{I}}{\sum y^2\frac{\delta s}{I}}$$

and the derivation is carried out by simple summation, when the values of the terms y and I can be measured independently for several selected points on the rib. This method is more approximate than the method of integration, but is sufficiently accurate for practical construction.

M   is the bending moment at any point when the span of the arch is considered as a simply supported beam under the given loading.

y   is the height of the arch at any point.

I   is the moment of inertia of the arch section at any point.

ds   is an elemental length measured along the curve of the arch rib.

δs   is a short length of the arch rib.

E   is presumed to be constant and has been cancelled from the expression.

Two-hinged Segmental Arches

4.1 *A segmental arch (i.e. one whose outline is a portion of a circle) of uniform moment of inertia has a span of* 60 *ft. and subtends* 90° *at the centre. One half of the arch is loaded with a uniformly distributed load of* 2 *tons per ft. run.*

(1) Find the radius of the arch.

From the properties of the circle (Fig. **4.1**),

$$h(60+h)=30^2,$$

whence
$$h=12\cdot42\,ft.$$

The radius is therefore $42\cdot42\,ft.$

(2) Determine the reactions and the expressions for bending moment, considering the span as a simply supported beam.

$$V_L=45\ tons. \quad V_R=15\ tons.$$

It is important to notice that there must be a separate bending moment expression for each portion of the span, the number of expressions depending on the nature of the loading. In the present problem two expressions are necessary, one for the left side and one for the right.

Left Side.

$$M=V_L(30-R\sin\,\theta)-\frac{w(30-R\sin\,\theta)^2}{2}$$
$$=450+15R\sin\,\theta-R^2\sin^2\,\theta\ ft.\text{-}tons.$$

Right Side.

$$M=V_R(30-R\sin\,\theta)$$
$$=450-15R\sin\,\theta\ ft.\text{-}tons.$$

(3) Since I is constant, the equation for H becomes

$$H=\frac{\int Myds}{\int y^2ds}.$$

M has been evaluated in (2), and from Fig. **4.1**,

$$y=R\cos\,\theta-30 \text{ and } ds=Rd\theta.$$

(4) These values can now be substituted in the expression for H, noting that the integration must be carried over the whole span for both numerator and denominator. It is usually better to carry out each portion of the integration in turn, and to determine H as a separate operation. In evaluating the definite integral it should be remembered that when $\theta=0$, $\cos\,\theta$ and $\cos 2\theta$ equal unity.

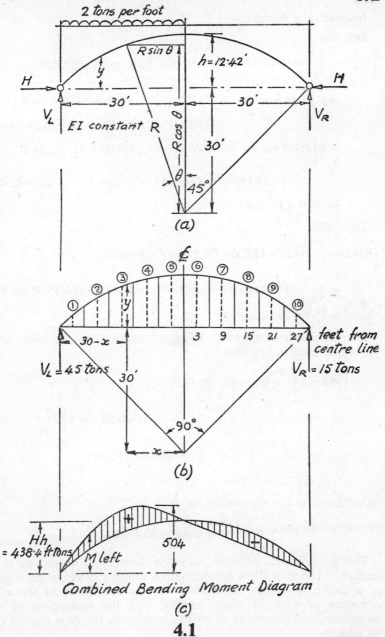

(a)

(b)

Hh
= 438·4 ft tons

M left

504

Combined Bending Moment Diagram

(c)

4.1

Integration of Numerator.

Left Side.

$$\int Myds = \int_0^{\pi/4} (450 + 15R\sin\theta - R^2\sin^2\theta)(R\cos\theta - 30)Rd\theta$$

$$= R\int_0^{\pi/4} (450R\cos\theta + 7{\cdot}5R^2\sin 2\theta - 13{,}500 - 450R\sin\theta$$

$$+ 15R^2 - 15R^2\cos 2\theta)d\theta - R^3\sin 2\theta(d\sin\theta)$$

$$= R\Big[450R\sin\theta - 3{\cdot}75R^2\cos 2\theta - 13{,}500\theta + 450R\cos\theta$$

$$+ 15R^2\theta - 7{\cdot}5R^2\sin 2\theta - \frac{R^3}{3}\sin^3\theta \Big]_0^{\pi/4}. \quad (R = 42{\cdot}42')$$

$$= 115{\cdot}0 \times 10^3 \ tons\text{-}ft.^3$$

Right Side.

$$\int Myds = \int_0^{\pi/4} (450 - 15R\sin\theta)(R\cos\theta - 30)Rd\theta$$

$$= R\int_0^{\pi/4} (450R\cos\theta - 15R^2\sin\theta\cos\theta - 13{,}500 + 450R\sin\theta)d\theta$$

$$= 75{\cdot}3 \times 10^3 \ tons\text{-}ft.^3$$

Integration of Denominator. (This integration can be taken over the whole span in one step.)

$$\int y^2 ds = 2\int_0^{\pi/4} (R\cos\theta - 30)^2 Rd\theta$$

$$= 2R\int_0^{\pi/4} \Big(\frac{R^2}{2} + \frac{R^2}{2}\cos 2\theta + 900 - 60R\cos\theta \Big)d\theta$$

$$= 2R\Big[\frac{R^2}{2}\theta + \frac{R^2}{4}\sin 2\theta + 900\theta - 60R\sin\theta \Big]_0^{\pi/4}$$

$$= 5{\cdot}4 \times 10^3 \ ft.^3$$

Substituting in the expression for H,

$$H = \frac{115{\cdot}0 + 75{\cdot}3}{5{\cdot}4} = 35{\cdot}2 \ tons.$$

This type of arch, however, is often more easily solved by summation. The arch rib is divided into a number of equal parts each of a length δs. The values of y and M are found at the mid-ordinates of each of these lengths, and the evaluation of the numerator and denominator of the expression for H is carried out by addition.

In (b) the span of the arch under consideration is seen to be divided into 10 equal parts. The approximation of dividing the span rather than the length of the arch rib is one which is permissible, provided that the arch is of a "flat" shape and the number of divisions is sufficiently large.

$$\delta s = \frac{\text{circumference}}{4} \times \frac{1}{10} = 6 \cdot 67 \, ft.$$

y is found from $(y+30)^2 + x^2 = 42 \cdot 42^2$.
$M_{\text{left}} = 45(30-x) - (30-x)^2 \, ft.\text{-}tons.$ $M_{\text{right}} = 15(30-x) \, ft.\text{-}tons.$
The values of M and y for each of the sections are tabulated as shown below.

Section	x (ft.)	y (ft.)	y^2	M (ft.-tons)	My (tons-ft.²)
1	27	2·8	7·8	126	352
2	21	6·9	47·5	324	2,235
3	15	9·8	96·0	450	4,410
4	9	11·5	132·0	504	5,800
5	3	12·3	151·0	486	5,980
6	3	12·3	151·0	405	4,980
7	9	11·5	132·0	315	3,620
8	15	9·8	96·0	225	2,210
9	21	6·9	47·5	135	932
10	27	2·8	7·8	45	126
Totals . .	—	—	868·6	—	30,645

$$H = \frac{\Sigma My \delta s}{\Sigma y^2 \delta s} = \frac{30,645 \times 6 \cdot 67}{868 \cdot 6 \times 6 \cdot 67} = 35 \cdot 3 \, tons.$$

The determination of the value of H is the first step to drawing the bending moment diagram, upon the dimensions of which the proportions of the arch section depend.

The negative bending moment caused by H at any point on the arch is $-Hy$, and since this term varies directly with y the negative bending moment diagram is similar to the shape of the arch itself.

First: draw the positive bending moment diagram caused by the loading. This can be done from the values given for M in the above table.

Second: determine the negative bending moment at the crown, caused by H. This $= Hh = 35 \cdot 3 \times 12 \cdot 42 = 438 \cdot 4 \, ft.\text{-}tons.$

From these data the combined bending moment diagram can be drawn as is shown in (c).

Exercises

A segmental two-hinged arch is of 56·5 ft. span and uniform I. Its radius is 40 ft. A concentrated load of 4 tons acts at 8 ft. 3 in. from the left hinge (measured horizontally). Draw the B.M.D. $(H=1\cdot65$ *tons.*)

Similar arch: span, 30 ft.; radius, 17 ft.; 12 tons at 10 ft. from left hinge.
$(H=6\cdot47$ *tons.*)

Varying Moment of Inertia

In Problem **4.1** the moment of inertia of the arch rib was assumed to be constant in order to simplify the working of the example. The moment of inertia of an arch, however, usually varies from a maximum at each abutment to a minimum at the crown. An expression which can be integrated in terms of s must therefore be obtained for I or the value of I must be known for each of the sections considered in the summation method.

Since such data concerning I cannot be obtained until the arch is designed the assumption may be made, during preliminary investigations, that the moment of inertia of the rib section varies directly as the secant of the angle of inclination of the arch. Such a relationship does not often occur in practice, but it forms a convenient and sufficiently exact assumption for the first rough estimation of the shape of the rib.

This assumption has the effect of simplifying the integration, as can be appreciated from the following:

It can be seen from Fig. **4.2a** that $ds=dx \sec \alpha$ where α is the slope of the arch rib, and from the assumption under discussion $I_x=I_c \sec \alpha$ where I_c is the moment of inertia of the arch rib at the crown. The expression for H then becomes

$$H=\dfrac{\displaystyle\int M\dfrac{ydx \sec \alpha}{I_c \sec \alpha}}{\displaystyle\int \dfrac{y^2dx \sec \alpha}{I_c \sec \alpha}}$$

and since I_c is constant

$$H=\dfrac{\displaystyle\int Mydx}{\displaystyle\int y^2dx}$$

For such an arch, therefore, the integration may be taken along the horizontal span instead of round the curve of the arch. This makes it convenient to derive the value of H for a segmental arch by using cartesian instead of polar co-ordinates.

74

4.2 *A circular arch of 40 ft. span and 10 ft. rise carries a 10-ton load at each quarter point. The moment of inertia of the cross-section varies directly as the secant of the slope of the arch rib. Determine H and draw the bending moment diagram for the arch.*

(1) Find the radius of the arch and the relation between x and y.

$$10(2R-10)=400.$$
$$R=25 \, ft.$$

From (a) $x^2+n^2=25^2.$
$$n=(25^2-x^2)^{1/2}.$$
$$\therefore \; y=(25^2-x^2)^{1/2}-15$$

also $y^2=850-x^2-30(25^2-x^2)^{1/2}.$

(2) In the integrations, the term $\int(25^2-x^2)^{1/2}dx$ appears. It is advisable to integrate this separately. Substituting
$$x=25 \sin \beta, \; dx=25 \cos \beta d\beta,$$

we have

$$\int(25^2-x^2)^{1/2} \, dx=25^2\int \cos^2 \beta d\beta=312\cdot5\int(1+\cos 2\beta)d\beta$$

$$=312\cdot5(\beta+\frac{1}{2}\sin 2\beta)=312\cdot5\left(\sin^{-1}\frac{x}{25}\right)+\frac{x}{2}(25^2-x^2)^{1/2}.$$

Between the limits $x=0$ to $x=10$, this becomes $=244$.

,,   ,, $x=0$ to $x=20$,,   ,, $=440$.

(3) *Integration of Numerator.* $\int Mydx.$

Sections AB and DC.

$$\int Mydx=2\int_{10}^{20}10(20-x)\{(25^2-x^2)^{1/2}-15\}dx=+5860 \; tons\text{-}ft.^3$$

Section BC.

$$\int Mydx=2\int_{0}^{10}100\{(25^2-x^2)^{1/2}-15\}dx=+18,800 \; tons\text{-}ft.^3$$

Total $\int Mydx=24,660 \; tons\text{-}ft.^3$

(4) *Integration of Denominator.* $\int y^2dx.$

$$\int y^2dx=2\int_{0}^{20}\{850-x^2-30(25^2-x^2)^{1/2}\}dx=2266 \; ft.^3$$

From these data $H=\dfrac{24,660}{2266}=10\cdot9 \; tons.$

The maximum negative bending moment (at the centre) $-10\cdot9 \times 10 \; ft.\text{-}tons$ and from this the bending moment diagram can be drawn as is shown in (b). Such an arch is probably more

4.2

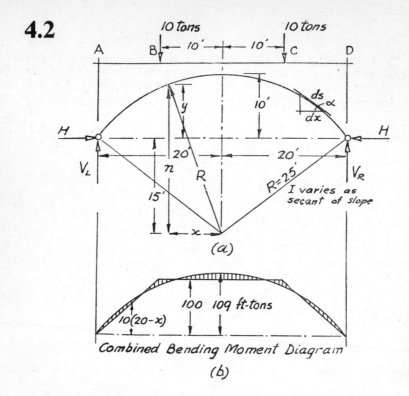

(a)

Combined Bending Moment Diagram

(b)

readily solved by the summation method and the student should check the above result in this way.

Parabolic Arches

In the following problems, which deal with arches of a parabolic shape, the assumption is made that the moment of inertia of the arch rib varies as the secant of the slope. The expression for H is therefore

$$H = \frac{\int M y \, dx}{\int y^2 \, dx}$$

and, as in Problem **4.2**, the integration can be taken horizontally along the span length.

M is found as before, by considering a simply supported beam of the same span as the arch and carrying the same loads.

y is found in terms of x from the equation to a parabola. In Fig. **4.3a**, if the origin of co-ordinates is taken at the abutment (A or B) the equation to the parabola is

$$y = Cx(l - x)$$

where C is a constant whose value depends on the proportions of the arch.

76

4.3 *A parabolic arch of 60 ft. span carries an 8-ton load at each third point. The rise of the arch is 6 ft., and the moment of inertia varies as the secant of the slope of the arch. Find the value of H and draw the bending moment diagram.*

This arch carries two symmetrically placed loads as did the arch in Problem **4.2**. The evaluation of H for such an arch can be simplified by carrying out the calculation with only one of the loads in place. Then from the principle of superposition the value of H so obtained is half of the value found when both loads are acting.

(1) Determine the value of C in the equation $y = Cx(l-x)$.

$$\text{When} \quad x = 30', \ y = 6'.$$
$$\therefore \ C = 6/900 = 1/150$$
$$\text{and} \quad y = \frac{x}{150}(60 - x)$$

is the equation to the parabola.

(2) Find the reactions and M when only the left-hand load is acting (b).

$$V_L = \frac{16}{3} \text{ tons.} \quad M \text{ (from } A \text{ to } C) = \frac{16}{3}x \text{ ft.-tons.}$$

$$V_R = \frac{8}{3} \text{ tons.} \quad M \text{ (from } B \text{ to } C) = \frac{8}{3}x \text{ ft.-tons.}$$

(3) *Integration of Numerator.* $\int Mydx.$

Section A to left-hand load.

$$\int_0^{20} \frac{16}{3}x \cdot \frac{x}{150}(60-x)dx = \frac{16 \times 20^3}{450}(20-5)$$
$$= 4 \cdot 3 \times 10^3 \text{ tons-ft.}^3$$

Section B to left-hand load.

$$\int_0^{40} \frac{8}{3}x \cdot \frac{x}{150}(60-x)dx = \frac{8 \times 40^3}{450}(20-10)$$
$$= 11 \cdot 4 \times 10^8 \text{ tons-ft.}^3$$

(4) *Integration of Denominator.*

$$\int_0^{60} \frac{x^2}{150^2}(60-x)^2dx = \frac{60^3}{150^2}\left(1200 - 1800 + \frac{60^2}{5}\right)$$
$$= 1 \cdot 15 \times 10^3 \text{ ft.}^3$$

From these data, $H = \dfrac{4 \cdot 3 + 11 \cdot 4}{1 \cdot 15} = 13 \cdot 6$ *tons* for one load

and 27·2 *tons* for both loads.

77

4.3

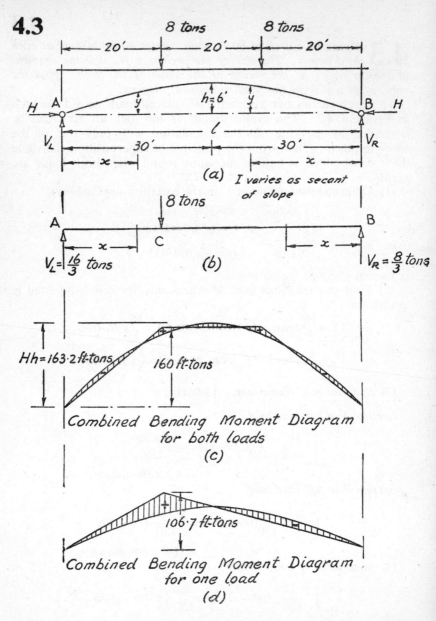

(a) *I varies as secant of slope*

$V_L = \frac{16}{3}$ tons   **(b)** $V_R = \frac{8}{3}$ tons

$Hh = 163 \cdot 2$ ft-tons   160 ft-tons

Combined Bending Moment Diagram for both loads
(c)

106·7 ft-tons

Combined Bending Moment Diagram for one load
(d)

Problems **4.2** and **4.3** show clearly how small is the bending moment in an arch when the load is symmetrically placed. If the load is uniformly distributed over a parabolic arch, there is no bending moment at any point of the arch rib.

4.4 *A uniformly distributed load of 4 tons per foot covers a 20-ft. length of the span of a parabolic arch as is shown in Fig. 4.4. The moment of inertia of the section varies as the secant of the slope of the arch and the span and rise are, respectively, 50 ft. and 15 ft. Draw the bending moment diagram for the arch, and determine the position and magnitude of the maximum bending moment.*

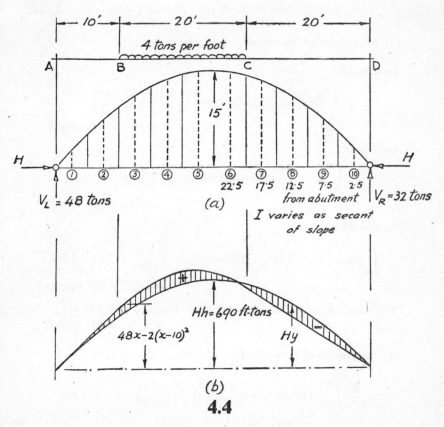

(a)

I varies as secant of slope

$V_L = 48$ tons

22·5 17·5 12·5 7·5 2·5
from abutment

$V_R = 32$ tons

$Hh = 690$ ft·tons

$48x - 2(x-10)^2$

Hy

(b)

4.4

(1) Determine the value of C in the equation $y = Cx(l-x)$.

     When $x = 25'$, $y = 15'$.

$$\therefore C = \frac{15}{25^2} = \frac{3}{125}.$$

$$\therefore y = \frac{3x}{125}(50-x) \text{ is the equation to the parabola.}$$

79

4.4

(2) Find the reactions and the equations for M when the arch is considered as a simply supported beam of 50 ft. span.

$$V_L = 1/50(80 \times 30) = 48 \; tons. \qquad V_R = 80 - 48 = 32 \; tons.$$

A to B: $\quad x = 0$ to $10 \; ft.$ $\quad M = 48x \; ft.\text{-}tons.$

D to C: $\quad x = 0$ to $20 \; ft.$ $\quad M = 32x \; ft.\text{-}tons.$

B to C: $\quad x = 10$ to $30 \; ft.$ $\quad M = 48x - 2(x-10)^2 \; ft.\text{-}tons.$

(3) *Integration of Numerator.* $\int M y \, dx.$

Section AB.

$$\int_0^{10} 48x \cdot \frac{3x}{125}(50-x)dx = \frac{144 \times 10^3}{125}\left(\frac{50}{3} - \frac{10}{4}\right)$$
$$= 16{\cdot}3 \times 10^3 \; tons\text{-}ft.^3$$

Section DC.

$$\int_0^{20} 32x \cdot \frac{3x}{125}(50-x)dx = \frac{96 \times 20^3}{125}\left(\frac{50}{3} - \frac{20}{4}\right)$$
$$= 71{\cdot}7 \times 10^3 \; tons\text{-}ft.^3$$

Section BC.

$$\int_{10}^{30}\{48x - 2(x-10)^2\}\frac{3x}{125}(50-x)dx = \frac{12 \times 10^4}{125}(123{\cdot}6 - 104)$$
$$= 188{\cdot}2 \times 10^3 \; tons\text{-}ft.^3$$

The total value of $\int M y \, dx$ is $16{,}300 + 71{,}700 + 188{,}200$
$$= 276{\cdot}2 \times 10^3 \; tons\text{-}ft.^3$$

(4) *Integration of Denominator.* $\int y^2 dx.$

$$\frac{9}{125^2}\int_0^{50} x^2(50-x)^2 dx = \frac{9 \times 50^5}{125^2}\left(\frac{1}{3} + \frac{1}{5} - \frac{1}{2}\right) = 6 \times 10^3 \; ft.^8$$

The value of $H = \dfrac{276}{6} = 46 \; tons$ and the negative bending moment

at the centre of the span

$$= -H \times 15 = -690 \; ft.\text{-}tons.$$

This type of problem can, of course, be solved by summation, but the integration method is generally more convenient.

Solution by summation method. (a) shows the span divided into ten equal parts, and the mid ordinates of these parts drawn as dotted lines. The values of M and y are found for each of the sections (1) to (10) and the results are tabulated below.

Section	x (ft.)	y (ft.)	M (ft.-tons)	y^2	My	$\delta x = \dfrac{50}{10} = 5'$
1	2·5 from A	2·85	120	8·1	344	
2	7·5 ,,	7·65	360	58·5	2,750	
3	12·5 ,,	11·25	587	126·7	6,600	
4	17·5 ,,	13·65	727	186·3	9,930	
5	22·5 ,,	14·85	767	220·6	11,400	
6	27·5 ,,	14·85	707	220·6	10,500	
7	17·5 from D	13·65	560	186·3	7,650	
8	12·5 ,,	11·25	400	126·7	4,500	
9	7·5 ,,	7·65	240	58·5	1,835	
10	2·5 ,,	2·85	80	8·1	228	
Totals .	—	—	—	1200·4	55,737	

$$H = \frac{\Sigma M y \delta x}{\Sigma y^2 \delta x} = \frac{55737 \times 5}{1200 \cdot 4 \times 5} = 46 \cdot 3 \ tons.$$

(b) shows the combined B.M.D.

Exercise

Check, by the *Strain Energy* method of Chapter 4, the value of H given in problem 4.5.

Thrust and Shear in a Two-hinged Arch

For purposes of design it is necessary to know not only the bending moment at any point on the arch rib but also the direct thrust and the shear at any section. Due to the presence of the horizontal reaction H at the abutments there is, on any section of the rib, a resultant force which may be resolved into two components:

    (*a*) a thrust parallel to the tangent to the arch rib at the section, and

    (*b*) a shear at right angles to the thrust.

To find the values of these quantities, the value of the horizontal reaction must first be found and then all the forces on one side of the section resolved into the two directions defined in (*a*) and (*b*).

6   81

4.5 The problem here is to find the value of the thrust and shear on a section 15 *ft.* from the left abutment, if the horizontal force H has been found to be 2·3 *tons* and the arch is of parabolic shape.

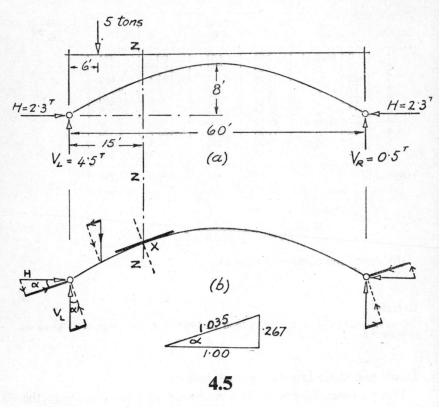

4.5

(1) *Slope of the arch rib* at the section in question.
Equation to the arch, the origin being at the left hinge, is

$$y = Cx(l - x).$$

Therefore $dy/dx = C(l - 2x) = 2/225(60 - 2x)$

and when $x = 15$ ft., $dy/dx = 0.267$.

This value represents the tangent of the angle of inclination of the arch rib at 15 ft. from the end. Later, the sine and cosine of this angle will be required. These are best obtained from a sketched diagram such as (b), the value of the hypotenuse being found from

82

the known lengths of the other two sides. This is a sufficiently accurate method, and is quicker than reference to mathematical tables.

$$\sin \alpha = \frac{0 \cdot 267}{1 \cdot 035} = 0 \cdot 258. \quad \cos \alpha = \frac{1 \cdot 000}{1 \cdot 035} = 0 \cdot 966.$$

(b) shows how the forces acting on the arch rib may be resolved into directions parallel to and at right angles to the slope of the rib at X. The components in these directions are shown in dotted and full lines, but the student will find the use of coloured pencils helpful in such problems. No "formula" should be memorised, but the arch ribs should be sketched for every solution.

(2) *Thrust at Section X* may be found by working from either side of the section, the forces, *on one side only*, being considered. Since the nett force in an arch rib is compressive, a positive sign is given to compressive and a negative sign to tensile forces. The values of the *full* lines are summed algebraically.

From the left side of section X

Thrust $= H \cos \alpha + V_L \sin \alpha - 5 \sin \alpha$

$\qquad = 2 \cdot 3 \times 0 \cdot 966 + 4 \cdot 5 \times 0 \cdot 258 - 5 \times 0 \cdot 258$

$\qquad = 2 \cdot 09$ *tons compression.*

From the right side of section X

Thrust $= H \cos \alpha - V_R \sin \alpha$

$\qquad = 2 \cdot 09$ *tons compression.*

(3) *Shear Force at Section X.* The convention of signs for shear force is given in Fig. **2.02** (Chapter 2, page 5). This time the algebraic sum of the *dotted* components is found on either side of the section.

From the left side of section X

Shear $= H \sin \alpha - V_L \cos \alpha + 5 \cos \alpha$

$\qquad = 1 \cdot 08$ *tons positive shear.*

A further discussion of thrust and shear in arches is to be found in Chapter 8, where the influence lines are derived.

Exercise

Find Thrust and Shear at quarter span in problems 4.3 and 4.4.

Temperature Thrust in a Two-hinged Arch

Up to this point the problems of Chapter 4 have dealt only with the bending moment, thrust and shear brought into play, at any section, by the loads on the arch and by the induced horizontal thrust at the hinges. Since, however, a two-hinged arch is fixed in position at the ends and lateral movement of the abutments is prevented, any rise or fall in temperature causes an alteration in the stresses in the rib. Such "temperature stresses" are often of importance in the design of reinforced concrete and steel arches and must be estimated for the probable range of temperature which will be experienced above and below the temperature at which the arch is built.

The immediate effect of a change in temperature is an alteration in the value of the horizontal thrust H. The additional horizontal reaction caused by a rise of temperature is given by

$$H_t = \frac{\alpha t l}{\int_0^l \frac{y^2 ds}{EI}}$$

where α is the coefficient of expansion,

$\quad t$ is the change in temperature in degrees F.

$\quad l$ is the span of the arch.

Secondary Stress in a Two-hinged Arch due to Rib Shortening

The material of which the rib is composed is compressible and the force H, induced by external loading and temperature change, causes a shortening of the arch length. Such deformation has the effect of relieving the rib of a small part of the original force H. Since the final value of H is less than that calculated for external loading when the rib is considered incompressible, the denominator of the expression for H must be increased to account for the effect of rib shortening. The value of H for external loading and rib shortening is

$$H = \frac{\int \frac{M y ds}{EI}}{\int y^2 \frac{ds}{EI} + \int \frac{ds}{AE}}$$

where A is the cross-sectional area of the rib at any point.

4.6 *Find the effect on the bending moment diagram of the arch in Problem* **4.1**, *of an increase in temperature of* 15° *F. The arch is of steel, whose coefficient of expansion is* 0·000006 *per* °*F.* I=4000 in.[4] *(uniform over whole arch).*

In Problem **4.1** the value of $\int y^2 ds$ was found to be $5\cdot4 \times 10^3$ *ft.*

Since *EI* was constant, and appeared in both numerator and denominator, its value was not required. Now, however, *EI* must be evaluated.

$$EI = 13,000 \times 144 \times \frac{4000}{12^4} = 361 \times 10^3 \; tons\text{-}ft.^2$$

$$\alpha t l = 0\cdot000006 \times 15 \times 60 = 5\cdot4 \times 10^{-3} \; ft.$$

$$H_t = \frac{361 \times 5\cdot4}{5\cdot4 \times 10^3} = 0\cdot36 \; tons.$$

Thus a temperature rise of 15° F. increases the negative bending moment by about 1 per cent. in this particular case.

Negative bending moment applied at the crown by *H*

$$= -35\cdot2 \times 12\cdot42 \; ft.\text{-}tons.$$

Negative bending moment applied at the crown by H_t

$$= -0\cdot36 \times 12\cdot42 \; ft.\text{-}tons.$$

$$\text{Total} = -35\cdot56 \times 12\cdot42 = -442 \; ft.\text{-}tons.$$

The positive moment at the crown is $+450$ *ft.-tons.*

$$\text{Nett moment} = +8 \; ft.\text{-}tons \; \text{at the crown.}$$

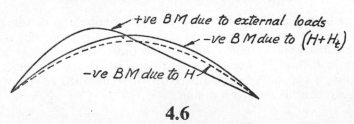

4.6

In **4.6** the original negative bending moment induced by the loading is shown dotted, and the negative bending moment induced by combined loading and increase in temperature is shown in full line. The diagram is not strictly to scale, but it can be seen that the effect (for this type of loading) of an increase in temperature is to increase the maximum negative bending moment and decrease the maximum positive bending moment.

4.7 The welded steel bridge rib is loaded as shown in (a). The rib cross-section consists of a web plate and two flange plates connected by fillet welds. The flanges are of uniform cross-sectional area and the web varies in depth.

For such a bridge, hinged at the abutments, the procedure in design is to determine, by some approximate method, the probable dimensions of the various sections, and then to check the values of the stresses at these sections. In this problem it is assumed that the preliminary design has been carried out, and that the stresses imposed by the given loading are required. Such a loading is only one of a variety of loadings whose effect must be determined before the bridge is deemed satisfactory.

Procedure. The centre line of the rib (not the span) is divided into 20 equal parts (Δs) and x and y determined for the mid-ordinate sections of these lengths. The table is made up as follows:—

Columns 1 and 2. x and y measured from a large scale drawing.

Column 3. Value of I for each section calculated from known dimensions of the cross-sections (not given here).

Column 5. M is the "free" bending moment at any section, considering the span as a simply supported beam under the given loading.

Columns 4 and 6. The value of H is obtained by dividing the sum of column 6 by the sum of column 4. It is to be noted that these sums were obtained with mixed units, the y and M terms being in ft. and lb. and the I terms in in.[4] Numerator and denominator can both be considered to be multiplied by 12, which (like Δs) cancels on division, and

$$H = \frac{31,740}{7 \cdot 28} \ lb. = 1 \cdot 95 \ tons.$$

Column 7. This gives the negative bending moment at each section, caused by H. The bending moment diagram can now be drawn by superimposing the "free" bending moment diagram on the diagram represented by Hy, which, as in previous problems, takes the form of the arch rib.

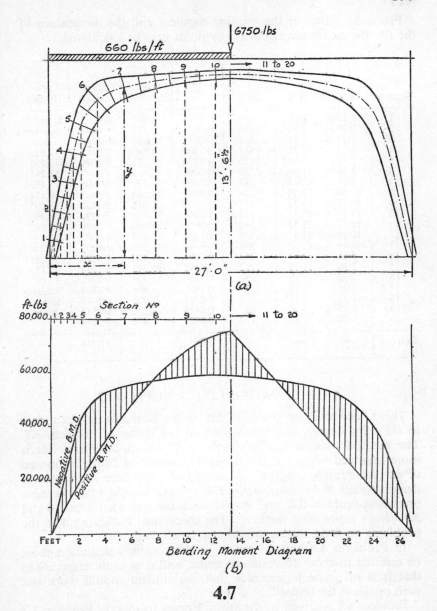

660 lbs/ft

6750 lbs

6 to 20 11 to 20

13' 6½"

27' 0"

(a)

ft-lbs
80,000
Section No
1 2 3 4 5 6 7 8 9 10 → 11 to 20

60,000

Negative B.M.D.

Positive B.M.D.

40,000

20,000

Feet 2 4 6 8 10 12 14 16 18 20 22 24 26

Bending Moment Diagram
(b)

4.7

87

4.7

From the values of the bending moment and the dimensions of the rib, the maximum stress at any point may be calculated.

	1	2	3	4	5	6	7
Section	x (ft.)	y (ft.)	I (in.^{4})	$\dfrac{y^2}{I}$	M (ft.-lb.)	My/I ÷1000	Hy (ft.-lb.)
1	0·25	1·10	157	0·0077	2,493	0·0175	Straight
2	0·75	3·30	256	0·0425	7,357	0·0948	line to
3	1·25	5·51	381	0·0797	12,056	0·1744	hinge
4	1·75	7·71	537	0·1107	16,590	0·2382	33,627
5	2·32	9·89	1030	0·0950	21,557	0·2070	43,135
6	3·56	11·76	1732	0·0798	31,623	0·2147	51,291
7	5·61	12·67	767	·02093	46,037	0·7605	55,260
8	7·86	13·04	294	0·5784	58,665	2·6020	56,874
9	10·10	13·32	157	1·1301	67,918	5·7622	58,095
10	12·37	13·52	140	1·3056	73,915	7·1381	58,968
11	12·37	13·52	140	1·3056	69,303	6·6926	58,095
12	10·10	13·32	156	1·1301	56,585	4·8007	56,874
13	7·86	13·04	294	0·5784	44,036	1·9532	55,260
14	5·61	12·67	767	0·2093	31,430	0·5192	51,291
15	3·56	11·76	1732	0·0798	19,945	0·1354	43,135
16	2·32	9·89	1030	0·0950	12,998	0·1248	33,627
17	1·75	7·71	537	0·1107	9,804	0·1408	Straight
18	1·25	5·51	381	0·0797	7,003	0·1013	line to
19	0·75	3·30	256	0·0425	4,202	0·0542	hinge
20	0·25	1·10	157	0·0077	1,401	0·0098	
Sum .	—	—	—	7·28	—	31,740	—

ARCHES WITH FIXED SUPPORTS

These "encastre" or "built-in" arches are statically indeterminate to the third degree, and correspond to the frame of Problem **3.7**. The method used in that problem could be applied here, but is more involved when the rib is curved instead of being composed of several straight lengths. A simplified procedure for arches is demonstrated in Problem **4.9**. The student should read the note preceding Problem **3.6**, and should work through Problems **3.6** and **3.7** before proceeding further. The theoretical background to the problems on fixed arches should also be studied.

In Problems **4.8** and **4.9** much of the routine calculation must be omitted in order to conserve space, and it is again emphasised that it is of prime importance that the student should work out each problem for himself.

Problem **4.8** is solved by the Strain Energy method of Problem **3.7** and Problem **4.9** shows how, for fixed arches, the work may be much reduced.

4.8 In this problem the "secant assumption" of Problem **4.8** is again adopted. Following the procedure of Problems **3.7** and **4.3** we have:

(1) *Equation to the Parabola,* with origin at one abutment, is

$$y=Cx(l-x)$$

when $x=30\,ft.$; $y=9\,ft.$; thus $C=0\cdot01$

and $y=\dfrac{x}{100}(60-x)=\dfrac{3}{5}x-\dfrac{x^2}{100}$

(2) *Table of Bending Moments.* As in Problem **3.7**, assuming that the forces and moments at the left abutment act in the direction shown at A in Fig. 3.7a.

	Bending Moment (M)	$\partial M/\partial H$	$\partial M/\partial V_L$	$\partial M/\partial M_L$	Limits
LB	$V_Lx+M_L-Hy-x^2/4$	$-y$	$+x$	$+1$	0–24
BR	$V_Lx+M_L-Hy-12(x-12)$	$-y$	$+x$	$+1$	24–60

First Equation.

$$\frac{1}{EI_c}\int_0^l M\frac{\partial M}{\partial H}dx=0$$

(since there is no relative horizontal movement of the abutments). I_c is the moment of inertia at the crown.

$$\frac{1}{EI_c}\int_0^{24}\left(-V_Lxy-M_Ly+Hy^2+\frac{x^2y}{4}\right)dx+\frac{1}{EI_c}\int_{24}^{60}\{-V_Lxy-M_Ly$$
$$+Hy^2+12y(x-12)\}dx=0.$$

It is evident that the integration for the V_L, M_L, and H terms may be taken over the whole span of 60 ft. in one operation. The equation then becomes

$$\int_0^{60}(-V_Lxy-M_Ly+Hy^2)dx+\int_0^{24}\frac{x^2y}{4}dx+\int_{24}^{60}12y(x-12)dx=0.$$

Before this can be evaluated y must be replaced by its value in terms of x (see (1) above). The work now becomes involved, but not more difficult than in Problem **3.7**. The student should evaluate the three terms of the integration in turn, and sum the results. Finally the first equation becomes

$$-10,800V_L-360M_L+2592H+81,174=0.$$

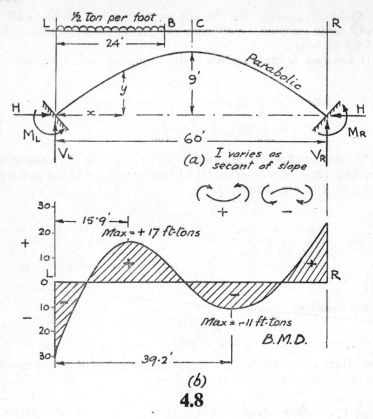

(a) *I varies as secant of slope*

(b)

4.8

Second Equation. Since there is no vertical settlement of the abutments

$$\frac{1}{EI_c}\int_0^l M\frac{\partial M}{\partial V_L}dx=0$$

and as for the first equation, taking the V_L, M_L, and H terms over the whole span in one operation, we have

$$\int_0^{60}(V_L x^2+M_L x-Hxy)dx+\int_0^{24}\left(-\frac{x^3}{4}\right)dx+\int_{24}^{60}\{-12x(x-12)\}dx=0.$$

Substituting for y and integrating,

$$+72,000V_L+1800M_L-10800H-611,712=0.$$

Third Equation. Since there is no rotation of the abutments

$$\frac{1}{EI_c}\int_0^l M\frac{\partial M}{\partial M_L}dx=0$$

and, taking the V_L, M_L, and H terms over the whole span in one operation, we have

$$\int_0^{60}(V_Lx+M_L-Hy)dx+\int_0^{24}\left(-\frac{x^2}{4}\right)dx+\int_{24}^{60}\{-12(x-12)\}dx=0.$$

$$+1800V_L+60M_L-360H-14,112=0.$$

(3) *Values of Support Reactions and Moment.* Simultaneous solution of the three equations, above, gives

$$V_L=+10\cdot5 \ tons; \quad H=+8\cdot1 \ tons; \quad M_L=-30\cdot1 \ ft.\text{-}tons.$$

The positive signs indicate that V_L and H act as shown in **3.7a** and **4.8a**. The negative sign for M_L indicates that the fixing moment does not act in the direction shown at A in Fig. **3.7a**, but in the opposite direction, as indicated in Fig. **4.8a**.

(4) *Maximum Bending Moment.* The maximum bending moment acting on the arch rib under the given loading conditions may be found either by calculation or by constructing a bending moment diagram, and finding the maximum ordinate by inspection.

(a) *By calculation:*

The bending moment at any section in the length LB is

$$M_x=10\cdot5x-30\cdot1-8\cdot1y-x^2/4.$$

Substituting for y and differentiating,

$$dM_x/dx=10\cdot5-8\cdot1(0\cdot6-0\cdot02x)-0\cdot5x.$$

When this is zero, the bending is a minimum, i.e. when $x=15\cdot9 \ ft.$ from left-hand end, $M_{max}=+17 \ ft.\text{-}tons.$

The bending moment at any section in the length BR is

$$M_x=10\cdot5x-30\cdot1-8\cdot1y-12x+144$$

$$dM_x/dx=10\cdot5-8\cdot1(0\cdot6-0\cdot02x)-12=0$$

whence $x-39\cdot2 \ ft.$ from left-hand end, $M_{max}=-11 \ ft.\text{-}tons.$

(b) *By constructing a bending moment diagram:*

Such a diagram is best constructed from ordinates obtained by preparation of a table, such as is shown below. The bending moment at all points on the arch rib due to the given loading is shown in Fig. **4.8b**.

4.8

x	y	$V_L x$	$-M_L$	$-Hy$	$-x^2/4$	$-12(x-12)$	B.M. (ft.-tons)
0	0	0	−30·1	0	0		−30·1
10	5	+105	−30·1	−40·5	−25		+9·4
20	8	+210	−30·1	−64·8	−100		+15·1
24	8·64	+252	−30·1	−70·0	−144		+7·9
30	9	+315	−30·1	−72·9		−216	−4·0
40	8	+420	−30·1	−64·8		−336	−10·9
50	5	+525	−30·1	−40·5		−456	−1·6
60	0	+630	−30·1	0		−576	+24·0

Revision Exercises

Find H and draw the B.M.D. for the two-hinged parabolic arches described. I varies with the secant of the slope of the rib.

Span	Rise	Loading	H
80 ft.	10 ft.	10 tons at crown	15.6 tons
76 ft.	10 ft.	2 tons/ft. over left half of span	72·0 tons
40 ft.	8 ft.	10 tons at crown and 10 tons at 10 ft. (horizontally) from left hinge	16·7 tons
40 ft.	10 ft.	12 tons at 10 ft. from left hinge	6·7 tons

The Elastic Centre

In a two-hinged arch problem the origin of co-ordinates may conveniently be taken at one of the abutments, but the problem just completed shows that such a choice for a fixed arch leads to involved work. Much of the laborious calculation encountered in Problem **4.8** can be avoided by a suitable choice of the origin of co-ordinates.

In the derivation of Problem **4.8** most of the involved calculation arises from terms containing $\int \dfrac{yx\,dx}{EI_c}$. If, therefore, for an arch such as that of Problems **4.8** and **4.9**, we take the YY axis through the crown, and the XX axis at such a height (z) above the springing line that $\int y\,dx = 0$, the work will be much facilitated.

4.9 Here again, as in the previous problem, the arch is of parabolic form, and the moment of inertia of the cross-section varies as the secant of the slope.

(1) *Equation to the Parabola.* In (a) the point 0 is the "elastic centre" and this new origin of co-ordinates is chosen so that $\int y dx$ is zero.

With C as origin the equation to the parabola is

$$y = Cx^2.$$

When $y = -15 \, ft.$ $x = 50 \, ft.$; thus $C = -3/500$.

With 0 as origin

$$y = (15 - z) - 3x^2/500.$$

(2) *Value of z.* z must be chosen so that $\int y dx$ is zero. Thus we have

$$\int y dx = 2 \int_0^{50} \left(15 - z - \frac{3x^2}{500} \right) dx$$

$$= 2 \left[15x - zx - \frac{x^3}{500} \right]_0^{50} = 0$$

$$z = 10 \, ft.$$

or the equation to the parabola with 0 as origin is

$$y = 5 - 3x^2/500.$$

Positive x is measured from 0 towards the right, and positive y upwards from 0.

(3) *Bending Moment Expressions.* In this problem the rib is cut at C, the crown. The left-hand and the right-hand halves then act as cantilevers, and the rib at C suffers a deflection and rotation. In order to bring C back to its original position two forces and a moment must be applied. Such forces and moments are shown applied at the elastic centre. H, V_O, and M_O are assumed in arbitrary directions, and their true directions will be shown by the signs of the values obtained at the end of the investigation.

The three equations necessary to obtain the values of H, V_O, and M_O are developed in a similar manner to that adopted in Problem **4.8**, this time working from the elastic centre as origin and integrating over the whole length of the arch rib. Because of the adoption of the "secant assumption" for I, the integration is made, as before, with respect to x.

	Bending Moment (M)	$\partial M/\partial H$	$\partial M/\partial V_O$	$\partial M/\partial M_O$	Limits
CL	$M_O - V_O(-x) - Hy - (-x)^2$	$-y$	$-x$	$+1$	$-50 - 0$
CB	$M_O - V_O x - Hy$	$-y$	$-x$	$+1$	$0 - 20$
BR	$M_O - V_O x - Hy - 20(x - 20)$	$-y$	$-x$	$+1$	$20 - 50$

93

4.9

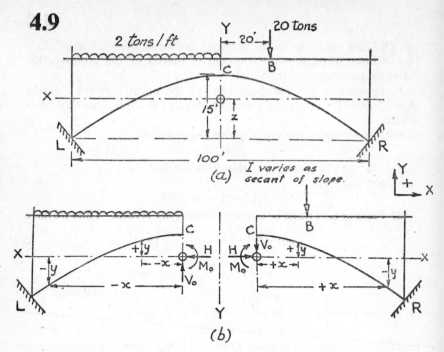

(a) *I varies as secant of slope.*

(b)

First Equation. Since there is no relative horizontal movement in the direction of H,

$$\frac{1}{EI_c}\int_{-50}^{+50} M\frac{\partial M}{\partial H}dx=0$$

Substituting for M, and $\partial M/\partial H$,

$$\int_{-50}^{0}(-M_0y-V_0xy+Hy^2+x^2y)dx+\int_{0}^{20}(-M_0y+V_0xy+Hy^2)dx$$

$$+\int_{20}^{50}\{-M_0y+V_0xy+Hy^2+20y(x-20)\}dx=0$$

But the origin of co-ordinates has been chosen so that terms involving $\int_{-50}^{+50} xdx$ and $\int_{-50}^{+50} ydx$ become zero. Also, as in Problem **4.8**, the H, V_0, and M_0 terms may be integrated in one step from -50 to $+50$. The simplified equation becomes

$$H\int_{-50}^{+50} y^2dx=\int_{-50}^{0}(-x^2)ydx+\int_{20}^{50}(-20x+400)ydx.$$

If we imagine C cut, and unsupported, the left- and right-hand portions of the arch rib act as cantilevers, and the bending produced is negative (convention of Chapter 2, Fig. **2.02**). In the left-hand

94

half the bending moment at a distance x from 0 is $(-x^2)$ and that at a distance x from C on the right-hand half of the beam is $-(20x-400)$. The above equation can thus be written:

$$H=+\frac{\left(\begin{array}{c}\text{Sum or integration of } y \text{ times the free bending moments}\\ \text{caused by both "left-hand" and "right-hand" loads}\end{array}\right)}{\int_{-50}^{+50}y^2dx}.$$

The advantage of taking the elastic centre as origin is now apparent, for the values of the statically indeterminate forces and moments can be found directly without the solution of simultaneous equations.

Each of the three terms required to find H should be evaluated separately, by substituting for y, as follows:

$$\int_{-50}^{0}(-x^2)ydx=-\int_{-50}^{0}\left(5x^2-\frac{3x^4}{500}\right)dx=+166\cdot67\times10^3 \ tons\text{-}ft.^3$$

$$-\int_{20}^{50}(20x-400)\left(5-\frac{3x^2}{500}\right)dx=+4\cdot41\times10^4 \ tons\text{-}ft.^3$$

$$\int_{-50}^{+50}y^2dx=\int_{-50}^{+50}\left(25+\frac{9x^4}{25\times10^4}-\frac{3x^2}{50}\right)dx=+2\times10^3 \ ft.^3$$

$$H=\frac{166\cdot67+44\cdot1}{2}=+105\cdot4 \ tons$$

Second Equation. Since there is no relative vertical movement,

$$\frac{1}{EI_c}\int_{-50}^{+50}M\frac{\partial M}{\partial V_o}dx=0$$

which, after substitution from the bending moment table, becomes

$$\int_{-50}^{0}(M_o+V_ox-Hy-x^2)(-x)dx+\int_{0}^{20}(-M_ox+V_ox^2+Hxy)dx$$

$$+\int_{20}^{50}\{-M_ox+V_ox^2+Hxy+20x(x-20)\}dx=0.$$

Eliminating terms containing $\int_{-50}^{+50}xdx$ and $\int_{-50}^{+50}ydx$, we have

$$V_o\int_{-50}^{+50}x^2dx=\int_{-50}^{0}(-x^2)xdx+\int_{20}^{50}(-20x+400)xdx.$$

Or, from a similar argument to that advanced for the first equation,

$$V_o=+\frac{\left(\begin{array}{c}\text{Sum or integration of } x \text{ times the free bending moments}\\ \text{caused by both "left-hand" and "right-hand" loads}\end{array}\right)}{\int_{-50}^{+50}x^2dx}$$

4.9

Evaluating the terms of this expression:

$$\int_{-50}^{0} -x^3 dx = +156{\cdot}25 \times 10^4 \ tons\text{-}ft.^3$$

$$-\int_{20}^{50} (20x^2 - 400x)dx = -36{\cdot}00 + 10^4 \ tons\text{-}ft.^3$$

$$\int_{-50}^{+50} x^2 dx = +8{\cdot}33 \times 10^4 \ ft.^3$$

$$V_O = +\frac{156{\cdot}25 - 36{\cdot}00}{8{\cdot}33} = +14{\cdot}4 \ tons.$$

Third Equation. Since there is no relative rotation,

$$\frac{1}{F.I_c}\int_{-50}^{+50} M\frac{\partial M}{\partial M_O} dx = 0$$

Proceeding as for the first and second equations we have

$$M_O\int_{-50}^{+50} dx = -\int_{-50}^{0} (-x^2)dx - \int_{20}^{50} (-20x + 400)dx$$

or
$$M_O = -\frac{\left(\begin{array}{c}\text{Sum or integration of the free bending moments}\\ \text{caused by "left-hand" and "right-hand" loads}\end{array}\right)}{\int_{-50}^{+50} dx}$$

Evaluating the terms of this expression,

$$\int_{-50}^{0} (-x^2)dx = -4{\cdot}17 \times 10^4 \ tons\text{-}ft.^3$$

$$\int_{20}^{50} (-20x + 400)dx = -0{\cdot}90 \times 10^4 \ tons\text{-}ft.^3$$

$$\int_{-50}^{50} dx = 100$$

$$M_O = -\frac{(-4{\cdot}17 - 0{\cdot}90) \times 10^4}{100} = +507 \ ft.\text{-}tons.$$

The positive signs for the three quantities H, V_O, and M_O indicate that the directions assumed (**4.9b**) are the correct ones. The final step of drawing the bending moment diagram and of determining stresses is carried out in the normal way.

The problems of Chapter 4 have illustrated the fundamental principles underlying the design of arches. When integration cannot be used the tabular method is adopted, and is more customary in practical design. Problem **8.8** gives an introduction to the tabular method for fixed arches. In the problems of Chapter 4, only one loading has been considered for each arch rib. In design, the worst possible combination of loads must be studied, and for this it is necessary to use influence lines. These are discussed in Chapter 8.

CHAPTER 5

METHOD OF SLOPE DEFLECTION

By this method it is possible to deal with beams under any degree of restraint at the ends, and with any settlement of the supports. The values of the moments at the ends of the beam depend on the loading, the angle through which the ends rotate, and the relative movement of the supports.

Slope Deflection Equations

In a beam AB (Fig. **5.0**).

M_{AB} represents the moment in the beam at A.

M_{BA}	,,	,,	moment in the beam at B.
θ_A	,,	,,	rotation of the end A after loading.
θ_B	,,	,,	rotation at the end B after loading.
K	,,	,,	moment of inertia divided by the length.
C_{AB}	,,	,,	fixed end moments at A due to the loading.
C_{BA}	,,	,,	fixed end moments at B due to the loading.
R	,,	,,	relative displacement of the end supports divided by the length of the beam.

The Slope Deflection Equations, which show the connection between these quantities and whose validity is proved in standard text-books on Theory of Structures, are as follows:

$$M_{AB}=2EK(2\theta_A+\theta_B-3R)-C_{AB}$$
$$M_{BA}=2EK(2\theta_B+\theta_A-3R)+C_{BA}.$$

Convention of Signs

θ is positive when the tangent to the beam turns in a clockwise direction.

R	,,	,,	beam rotates in a clockwise direction.
M	,,	,,	moment acts in a clockwise direction on the beam.

C is a numerical value and takes the sign shown in the equation.

97

7

5.0

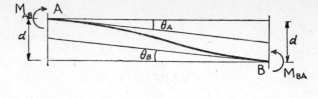

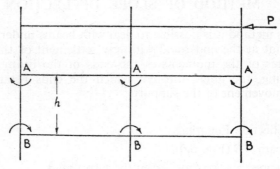

$$\frac{3M_A + 3M_B}{h} - P = 0$$

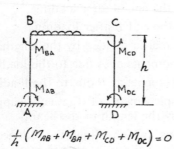

$$\frac{1}{h}\left(M_{AB} + M_{BA} + M_{CD} + M_{DC}\right) = 0$$

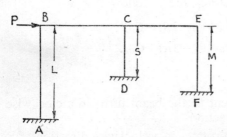

$$\frac{M_{AB} + M_{BA}}{S} + \frac{M_{CD}}{L} = 0$$

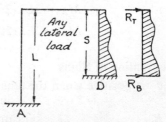

$$\frac{M_{AB} + M_{BA}}{L} + \frac{M_{CD} + M_{DC}}{S} + \frac{M_{EF} + M_{FE}}{M} + P = 0$$

$$\frac{M_{AB} + M_{BA}}{L} + \frac{M_{CD} + M_{DC}}{S} - R_T = 0$$

5.0

5.1 Since there is no relative settlement of the supports the term R is zero, and as the ends A and C are fixed, both θ_A and θ_C are also zero, there being no rotation at fixed ends.

The only unknown left, therefore, is θ_B and this illustrates how the use of the slope deflection method often results in a reduction in the number of unknowns. This problem, if solved by the methods of Chapter 2, contains three unknown quantities.

From the table of fixing moments given in the Appendix the values of C can be obtained.

Fixing moments for span AB are equal and $= wl/8 = 100/8$ *ft.-tons.*

Fixing moments for span BC are also equal and

$$= wl^2/12 = 100/12 \; \textit{ft.-tons.}$$

Equations. Substituting in the Slope Deflection equations

$$M_{BA} = 2EK(2\theta_B + \theta_A) + 100/8$$
$$M_{BC} = 2EK(2\theta_B + \theta_C) - 100/12.$$

But, for equilibrium

$$M_{BA} + M_{BC} = 0.$$

Also, K is the same for both spans, since each is 10 ft. long, and the moment of inertia is uniform along the beam.

$$\therefore \; 4EK\theta_B + 100/8 + 4EK\theta_B - 100/12 = 0$$
$$\theta_B = -\frac{0\cdot52}{EK}.$$

The moments at A, B, and C are now easily found by substituting the value of θ_B in the slope deflection equations as follows:

$$M_{AB} = 2EK(\theta_B) - 100/8 = -13\cdot54 \; \textit{ft.-tons.}$$
$$M_{BA} = -M_{BC} = 2EK(2\theta_B) + 100/8 = +10\cdot42 \; \textit{ft.-tons.}$$
$$M_{CB} = 2EK(\theta_B) + 100/12 = +7\cdot29 \; \textit{ft.-tons.}$$

Finally, the bending moment and shear force diagrams can be drawn as demonstrated in Chapter 2, but it should be noted that the convention of signs for bending moment is different.

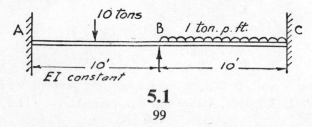

5.1

5.2 *When the beam shown in this figure is loaded, the support B sinks $\frac{1}{2}$ in. below A and support C sinks a further 1 in. below B.* As in Problem **5.1** the moment of inertia is constant along the length of the beam, and thus EK is the same for both AB and BC. The procedure is similar to that followed in Problem **5.1** but this time R is not zero. θ_A and θ_C are both zero.

Value of R and C. The value of R for both AB and BC is positive, for in both spans the beam rotates in a clockwise direction as a consequence of the sinking of the supports.

For AB $\quad R=\dfrac{0 \cdot 5}{12 \times 12}=\dfrac{1}{288}.$ $\qquad C=\dfrac{wl^2}{12}=12\,ft.\text{-}tons.$

For BC $\quad R=\dfrac{1}{12 \times 12}=\dfrac{1}{144}.$ $\qquad C=12\,ft.\text{-}tons.$

Equations. As in the preceding problem $M_{BA}+M_{BC}=$
$$M_{BA}=2EK(2\theta_B-3/288)+12$$
$$M_{BC}=2EK(2\theta_B-3/144)-12$$
$$4EK\theta_B-EK/48+12+4EK\theta_B-EK/24-12= \therefore$$
$$8EK\theta_B=EK/16$$
$$\theta_B=+1/128.$$

Moments. The moments are evaluated by substituting this value in the Slope Deflection equations, but exact results require a knowledge of E and K in ft.-ton units, which was not required when the supports remained rigid.

$$M_{AB}=2EK\left(\theta_B-\frac{3}{288}\right)-12=-\frac{EK}{192}-12\,ft.\text{-}tons.$$

$$M_{BA}=-M_{BC}=2EK\left(2\theta_B-\frac{3}{288}\right)+12=+\frac{EK}{96}+12\,ft.\text{-}tons.$$

$$M_{CB}=2EK\left(\theta_B-\frac{3}{144}\right)+12=-\frac{5EK}{192}+12\,ft.\text{-}tons.$$

If the beam is a B.S.B. 117 (10 in. $\times$ $4\frac{1}{2}$ in.), $I=122\ in.^4$
$$K=\frac{I}{l}=\frac{122}{12^4 \times 12}\,ft.^3$$
$$E=13{,}000 \times 144\ tons\ per\ sq.\ ft.$$
$$EK=\frac{13{,}000 \times 122}{1728}=917\,ft.\text{-}tons.$$

Substituting in the expressions for bending moment, we have,
$$M_{AB}=-16 \cdot 8\,ft.\text{-}tons;\ M_{BA}=+21 \cdot 6\,ft.\text{-}tons;\ M_{CB}=-11 \cdot 9\,ft.\text{-}tons.$$

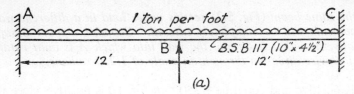

(a)

B.M.D. and Deflected Form when A and C sink

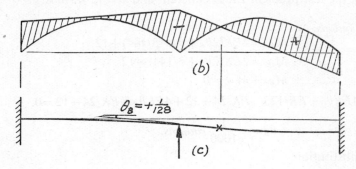

(b)

$\theta_B = +\frac{1}{128}$

(c)

B.M.D. and Deflected Form when A rotates and B sinks

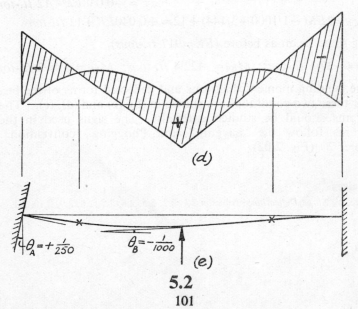

(d)

$\theta_A = +\frac{1}{250}$ $\theta_B = -\frac{1}{1000}$

(e)

5.2

If the same beam (Fig. 5.2) settles under load in a different way, the moments will be completely altered. Suppose support B sinks vertically through 1 in., while the wall into which A is built rotates in a clockwise direction through an angle of 1/250 radians.

θ_A is positive and equal to 1/250. R for AB is positive, since the beam rotates in a clockwise direction as a consequence of the sinking of the support. In BC, however, the value of R is negative. As in the last problem EK is constant and $C=12$ *ft.-tons.*

Equations.

$$M_{BA}=2EK(2\theta_B+\theta_A-3/144)+12$$
$$M_{BC}=2EK(2\theta_B+3/144)-12$$
$$M_{BA}+M_{BC}=0$$

$$4EK\theta_B+EK/125-EK/24+12+4EK\theta_B+EK/24-12=0.$$

$$\theta_B=-\frac{1}{1000} \text{ radians.}$$

By substitution,

$$M_{AB}=2EK(2/250-1/1000-3/144)-12=-0{\cdot}028EK-12 \text{ ft.-tons.}$$
$$M_{BA}=-M_{BC}=2EK(-2/1000+1/250-3/144)+12$$
$$=-0{\cdot}038EK+12 \text{ ft.-tons.}$$
$$M_{CB}=2EK(-1/1000+3/144)+12=+0{\cdot}040EK+12 \text{ ft.-tons.}$$

Using same beam as before ($EK=917$ *ft.-tons*).

$M_{AB}=-37{\cdot}4$ *ft.-tons*; $M_{BA}=-22{\cdot}4$ *ft.-tons*; $M_{BC}=+48{\cdot}3$ *ft.-tons.*

The bending moment diagrams and deflected forms of the beam under these two conditions are shown in (**b**), (**c**) and (**d**), (**e**). These diagrams should be studied carefully. The signs used in these diagrams follow the "sagging" and "hogging" convention of Chapter 2 (Fig. **2.02**).

Exercises

Solve, by *Slope Deflection*, problems 2.1, 2.2, 2.5, 2.6, 2.7, 2.8, 2.9, 2.10, 3.1.

5.3 Although in this beam the moment of inertia of the section is constant, the lengths of the spans are different and the values of K, therefore, are not equal. When the beam is loaded, the supports yield. B sinks 1 in., C sinks 2 in., and A and D remain at their original levels. Since all supports were originally at the same level, this means that the value of R for spans AB and BC is positive, and for span CD is negative, the rotation of the beam CD due to the sinking of C being in a counter-clockwise direction. The support at D remains level, but the wall into which A is built yields, turning through 1/200 radians in a clockwise direction.

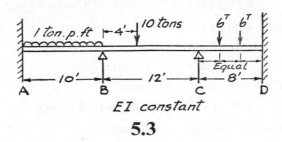

EI constant

5.3

The following table gives the values of the various terms of the Slope Deflection equations for the three spans, and the student is advised to check these figures, paying particular attention to the signs.

Span AB	Span BC	Span CD
$R = +\dfrac{1}{120}$	$R = +\dfrac{1}{144}$	$R = -\dfrac{1}{48}$
$K = \dfrac{I}{10}$	$K = \dfrac{I}{12}$	$K = \dfrac{I}{8}$
$\theta_A = +\dfrac{1}{200}$	θ_B and θ_C unknown	$\theta_D = 0$
$C_{AB} = \dfrac{25}{3}$	$C_{BC} = \dfrac{160}{9}$	$C_{CD} = \dfrac{32}{3}$
$C_{BA} = \dfrac{25}{3}$	$C_{CB} = \dfrac{80}{9}$	$C_{CD} = \dfrac{32}{3}$

As is shown in the table, there remain only two unknowns, namely, the angles through which the beam at B and at C rotates.

5.3

The determination of these values requires two equations which are obtained from the following conditions of equilibrium:

$$M_{BA}+M_{BC}=0.$$
$$M_{CB}+M_{CD}=0.$$

Equations for BA and BC.

$$M_{BA}=EI/5(2\theta_B+1/200-3/120)+25/3$$
$$M_{BC}=EI/6(2\theta_B+\theta_C-3/144)-160/9.$$

Summing these and equating to zero,

$$0.73\theta_B+0.17\theta_C-0.0075-9.47/EI=0.$$

Equations for CB and CD.

$$M_{CB}=EI/6(2\theta_C+\theta_B-3/144)+80/9$$
$$M_{CD}=EI/4(2\theta_C+3/48)-32/3.$$

Summing these and equating to zero,

$$0.17\theta_B+0.83\theta_C+0.012-1.78/EI=0.$$

Solving these two equations simultaneously,

$$\theta_C=-0.0174-0.54/EI$$
and
$$\theta_B=+0.0141+13.10/EI$$

By substitution in the original Slope Deflection equations the various moments may be obtained. For instance

$$M_{CD}=EI/4(-1.08/EI+1/16-0.035)-10.67$$
$$=-10.94+0.0069EI.$$

The fixing moments at A and D can be found in the same way by substituting the values of θ_B and θ_C in the Slope Deflection equations for AB and DC.

$$M_{AB}=EI/5(1/100+\theta_B-3/120)-25/3.$$
$$M_{DC}=EI/4(\theta_C+3/48)+32/3.$$

Draw the bending moment diagram and deflected form of the beam, choosing a suitable value for EI.

Exercises

Solve, by *Slope Deflection*, problems 3.2, 3.6, 6.1, 6.2, 6.3.

PORTAL AND BUILDING FRAMES

In Problems **5.1** to **5.3** the equations required for evaluating the statically indeterminate quantities have been obtained from conditions of equilibrium at each of the joints. The condition for equilibrium of a joint is that the sum of the moments in the members at that joint should be zero. For most portal and building frames, however, the number of equations obtained in this way is not sufficient to evaluate all the statically indeterminate quantities, and the additional conditions of equilibrium required are obtained from a consideration of the shear force exerted on the structure by external loading. When a frame carries a horizontal load (as in Problem **3.5**) the horizontal resistance exerted by the frame must be equal and opposite to the horizontal shear imposed by the loading.

The horizontal shear exerted by a member is equal to the sum of the moments at the top and bottom, divided by the length of the member. Hold a flexible scale *CD* vertically, and with finger and thumb apply a clockwise (positive) moment to the bottom (*D*). The upper end of the scale (*C*) will require a restraint to prevent its moving towards the right. If a similar clockwise moment is applied at the top (*C*) the bottom of the scale will have a tendency to move towards the left.

When both these moments are applied simultaneously the scale tends to "lean over" to the right, and will therefore resist the effect of a horizontal force towards the left at the top of the scale, and a similar force towards the right at the bottom of the scale. These two forces (*R*) form a couple on the length of the member.

If there are two or more vertical members resisting the horizontal force (as in portal and building frames) the horizontal shear *R* is found for each member, and the sum of these values is equivalent to the external horizontal force acting on the frame.

The examples given in **5.0** illustrate the types of relationship which are used in determining equations in the Slope Deflection method and in the method of Moment Distribution (Chapter 6).

5.4 Any eccentricity about the central vertical axis of a portal frame, whether it be unsymmetrical construction of the frame, unsymmetrical vertical loading or a horizontal load, causes the frame to sway sideways, so that the deflection of B relative to A and the deflection of C relative to D is not zero. These deflections, though unknown, are equal, for the bending of the beam BC under loading is so slight as to make no alteration in its length, and C must move laterally through the same distance as B.

In this frame the three unknown quantities are the angles of rotation of B and C (θ_B and θ_C) and the lateral deflection of the frame at the level of the beam ($R=d/12$). The stiffnesses of all three members are equal and C_{AB}, etc. $=0$ throughout.

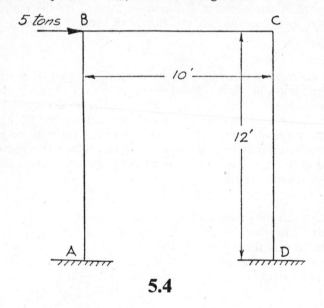

5.4

Three equations are required, and these are provided by the conditions of equilibrium.

$$M_{BA}+M_{BC}=0; \quad M_{CB}+M_{CD}=0$$

and

5×12 *ft.-tons* (clockwise moment) $+(M_{AB}+M_{BA})+(M_{CD}+M_{DC})$
(anti-clockwise moment) $=0$

or $\quad 5$ *tons* (to the right at the top) $+\dfrac{M_{AB}+M_{BA}}{12}+\dfrac{M_{CD}+M_{DC}}{12}$
(to the left at the top) $=0$.

First write down all the equations

$$M_{AB}=2EK(\theta_B-3R)$$
$$M_{BA}=2EK(2\theta_B-3R)$$
$$M_{BC}=2EK(2\theta_B+\theta_C)$$
$$M_{CB}=2EK(2\theta_C+\theta_B)$$
$$M_{CD}=2EK(2\theta_C-3R)$$
$$M_{DC}=2EK(\theta_C-3R).$$

Now, from above

$$M_{BA}+M_{BC}=0.$$
$$2\theta_B-3R+2\theta_B+\theta_C=0$$
$$4\theta_B+\theta_C-3R=0 \quad . \quad . \quad . \quad . \quad . \quad (1)$$

also $\qquad M_{CB}+M_{CD}=0$

$$2\theta_C-3R+2\theta_C+\theta_B=0$$
$$4\theta_C+\theta_B-3R=0 \quad . \quad . \quad . \quad . \quad .$$

and $\qquad M_{AB}+M_{BA}+M_{CD}+M_{DC}+60=0$

$$\theta_B+\theta_C-4R=-\frac{10}{KE} \quad . \quad . \quad . \quad .$$

(1) and (2) show that $\qquad\qquad \theta_B=\theta_C,$

whence (1) becomes $\qquad\qquad 5\theta_C-3R=0,$

and (3) becomes $\qquad\qquad 2\theta_C-4R=-\dfrac{10}{EK}$

Solving these two equations simultaneously

$$R=+\frac{25}{7EK} \quad \text{and} \quad \theta_B=\theta_C=+\frac{15}{7EK}$$

Substitute in the original Slope Deflection equations in order to determine the values of the various bending moments, and draw the bending moment diagram. The signs of the moments show whether the beam or column has been bent in a clockwise or counter-clockwise direction. It is very instructive to attempt in all such problems as this to draw the deflected form of the frame. Once practice is acquired, such sketches give a clear mental picture of what is taking place.

$$M_{DC}=M_{AB}=2EK(15/7EK-75/7EK)=-17\cdot15 \; ft.\text{-}tons.$$
$$M_{CD}=M_{BA}=2EK(30/7EK-75/7EK)=-12\cdot85 \; ft.\text{-}tons.$$

Note that M_{AB} is not one-half of M_{BA} as in Problem **3.12**, because of the lateral movement of B relative to A.

5.5 This problem illustrates the use of the "shear relationship" when the two members concerned are not parallel, as in Problem **5.4**, but are in one straight line.

Here two equations are required, in order to find the slope at B (θ_B) and the displacement at B (d). *Both of these are given positive signs*, and the result of the calculation will show whether this assumption is correct or not.

For equilibrium at B the applied 5 ft.-tons obviously causes negative moments in BA and BC. Thus,

First condition of equilibrium : $M_{BA}+M_{BC}=-5.$

Also, the horizontal forces brought into play at A and C **(c)** must be equal and opposite, since no other horizontal force acts on the pillar. BA must exert a shear force of P tons towards the right at the bottom, and BC must exert a shear force of P tons towards the left at the top. Thus,

$$M_{AB}+M_{BA}+5P=0; \quad M_{BC}+M_{CB}+10P=0.$$

Eliminating P we have:

Second condition of equilibrium : $2(M_{AB}+M_{BA})-(M_{BC}+M_{CB})=0.$

Writing down the values of the various moments from the Slope Deflection relationships, noting that d is positive for AB and negative for BC.

$$M_{AB}=2EI/5(\theta_B-3d/5) \quad : M_{BC}=2EI/10(2\theta_B+3d/10)$$
$$M_{BA}=2EI/5(2\theta_B-3d/5) \quad : M_{CB}=2EI/10(\theta_B+3d/10).$$

Substituting these values in the two equations of equilibrium and solving simultaneously,

$$\theta_B=-\frac{5\cdot55}{EI}; \quad d=-\frac{9\cdot26}{EI} \ ft.$$

Substituting the values of θ_B and d in the expressions for the bending moments,

$M_{AB}=0$;	$M_{BA}=-2\cdot22\ ft.\text{-}tons$;
$M_{BC}=-2\cdot78\ ft.\text{-}tons$;	$M_{CB}=-1\cdot67\ ft.\text{-}tons$

from which the bending moment and deflected form diagrams can be drawn. **(b)**, **(c)** The fixing moments are anti-clockwise and therefore negative. M_{AB} is zero because of the loading and proportions of the column.

5.6

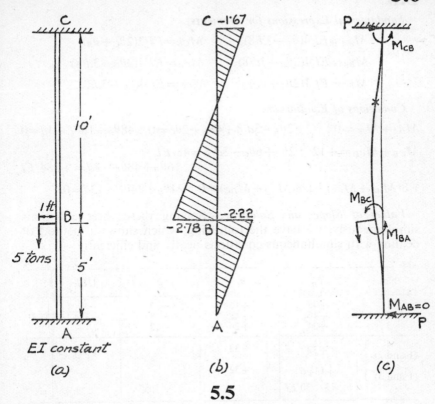

5.5

Exercises

Solve, by *Slope Deflection*, problems 6.4, 7.1, 7.2, 3.3, 7.3.

5.6 As another example of a frame carrying an externally applied moment, this problem shows a portal frame which is monolithic with a loaded cantilever. The procedure follows the same lines as in the previous problems of this chapter. The "unknowns" are θ_B, θ_C and d the sidesway of the frame. These are all given positive signs in the original equations.

Since the frame is not restrained in any way at the top, and has no horizontal load acting there, the internal shear in the two columns must be zero.

109

5.6

Fundamental Expressions for Moments.

$$M_{AB} = EI/4(\theta_B - 3d/8) \quad : \quad M_{CB} = EI/2(2\theta_C + \theta_B)$$

$$M_{BA} = EI/4(2\theta_B - 3d/8) \quad : \quad M_{CD} = EI/4(2\theta_C - 3d/8)$$

$$M_{BC} = EI/2(2\theta_B + \theta_C) \quad : \quad M_{DC} = EI/4(\theta_C - 3d/8).$$

Conditions of Equilibrium.

$$M_{BA} + M_{BC} = 0 \qquad : 2\theta_B - 3d/8 + 4\theta_B + 2\theta_C = 0 : 48\theta_B + 16\theta_C - 3d = 0$$

$$M_{CB} + M_{CD} = +12 \ : \ 2\theta_B + 6\theta_C - 3d/8 = 48/EI :$$
$$16\theta_B + 48\theta_C - 3d = +384/EI$$

$$1/8(M_{AB} + M_{BA}) + 1/8(M_{CD} + M_{DC}) = 0 : 24\theta_B + 24\theta_C - 12d = 0.$$

Values of Slopes and Sidesway. Solving these three equations simultaneously, we have the following, which shows a method of dealing with simultaneous equations neatly and efficiently:

No.	θ_B	θ_C	d	$1/EI$	
1	+48	+16	− 3	—	$\left.\begin{array}{c} \\ \\ \\ \end{array}\right\}=0$
2	+16	+48	− 3	−384	
3	+24	+24	−12	—	
4 (1 and 2)	+32	−32		+384	
5 (1 and 3)	+134·6 $\theta_B = -2·30/EI$	+32		—	
4	−2·30	+1 $\theta_C = +9·70/EI$	+12		
3	−4·60	+19·39	− 1 $d = +14·79/EI$		

Eliminating d between (1) and (2) and between (1) and (3) the equations (4) and (5) are obtained. θ_B is found by eliminating θ_C between (4) and (5), and the other values appear by substitution.

Substituting the values of θ_B, θ_C and d in the fundamental expressions for moment, the bending moment diagram and the deflected form can be drawn.

$$M_{AB} = -1·96 \ \textit{ft.-tons} \quad : \quad M_{CB} = +8·54 \ \textit{ft.-tons}.$$

$$M_{BA} = -2·54 \ \text{,,} \quad : \quad M_{CD} = +3·46 \ \text{,,}$$

$$M_{BC} = +2·54 \ \text{,,} \quad : \quad M_{DC} = +1·04 \ \text{,,}$$

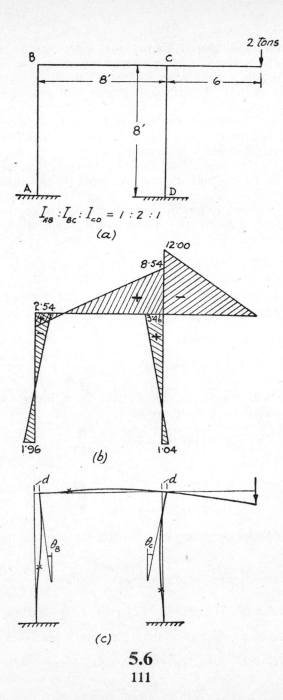

5.7 Two-hinged portal frames are more readily solved by methods given in other chapters than by the Method of Slope Deflection, but Problems **5.7** and **5.8** illustrate several important points and afford direct comparison with solutions by other methods.

Fundamental Expressions for Moments. Since the columns at A and D are not fixed in direction, θ_A and θ_D are unknown. There are five quantities whose magnitudes must be determined. They are θ_A, θ_B, θ_C, θ_D and d. θ_A and θ_D, however, can be obtained in terms of θ_B, θ_C and d, and the work is thus considerably shortened.

In this problem a further simplification can be obtained by dividing the lengths by 10. The following expressions can now be written down:

$$M_{AB}=2EI(2\theta_A+\theta_B-3d)=0 \qquad \therefore \ \theta_A=3d/2-\theta_B/2$$
$$M_{DC}=4EI/5(2\theta_D+\theta_C-6d/5)=0 \quad \therefore \ \theta_D=3d/5-\theta_C/2$$
$$M_{BA}=2EI(2\theta_B+\theta_A-3d) \qquad : \quad M_{CB}=4EI/5(2\theta_C+\theta_B)+P/4$$
$$M_{BC}=4EI/5(2\theta_B+\theta_C)-P/4 \qquad : \quad M_{CD}=4EI/5(2\theta_C+\theta_D-6d/5).$$

Conditions of Equilibrium.

$$M_{BA}+M_{BC}=0 \ : \ M_{CB}+M_{CD}=0 \ : \ \frac{M_{AB}+M_{BA}}{1}+\frac{M_{CD}+M_{DC}}{2\cdot5}=0.$$

Expanding and substituting for θ_A and θ_D in order to limit the unknown quantities to three, we have

$$4\cdot6\theta_B+0\cdot8\theta_C-3\cdot0d=+\frac{P}{4EI}$$

$$0\cdot8\theta_B+2\cdot8\theta_C-0\cdot5d=-\frac{P}{4EI}$$

$$3\cdot0\theta_B+0\cdot5\theta_C-3\cdot2d=0.$$

Values of Slopes and Sidesway. Solving the three equations simultaneously,

$$\theta_B=+\frac{0\cdot163P}{EI} \ : \ d=+\frac{0\cdot136P}{EI} \ : \ \theta_A=+\frac{0\cdot123P}{EI}.$$

But $M_{BA}=H\times1$. Thus $H=2(0\cdot324+0\cdot123-0\cdot408)P=0\cdot08P$.

Using the lengths shown in the figure, and a load of $P=4$ tons,

$$H=0\cdot32 \ tons \ : \ M_{BA}=3\cdot2 \ ft.\text{-}tons \ : \ M_{CD}=8\cdot0 \ ft.\text{-}tons.$$

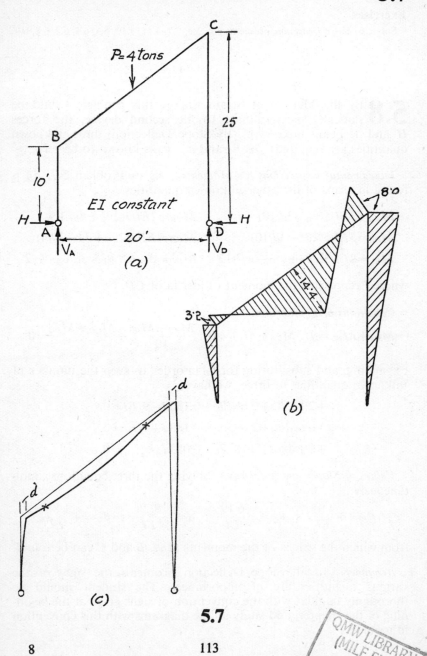

$P = 4\ tons$

25

B

$10'$

$EI\ constant$

H

A

$20'$

D

H

V_A

V_D

(a)

$8\cdot0$

$14\cdot4$

$3\cdot2$

(b)

d

d

(c)

5.7

Exercises

Solve, by *Slope Deflection*, problems 3.4, 3.5, 3.7, 3.11, 3.12, 6.5, 6.6, 6.7, 6.8, 6.9.

5.8 By the Method of Strain Energy this problem would be statically indeterminate to the second degree, the forces H and V being unknown. By Slope Deflection, three unknown quantities are required: θ_B, θ_C and d. θ_A is known to be zero.

Fundamental Expressions for Moments. As in Problem **5.7**, θ_D is found in terms of the other unknown quantities.

$$M_{AB}=3EI/5(\theta_B-3d/10) \qquad M_{CB}=2EI/3(2\theta_C+\theta_B)+9$$

$$M_{BA}=3EI/5(2\theta_B-3d/10) \qquad M_{CD}=EI/2(2\theta_C+\theta_D-3d/4)$$

$$M_{BC}=2EI/3(2\theta_B+\theta_C)-wl^2/12 \qquad M_{DC}=0. \quad \therefore \; \theta_D=3d/8-\theta_C/2$$

where I represents the moment of inertia of CD.

Conditions of Equilibrium.

$$M_{BA}+M_{BC}=0 \; : \; M_{CB}+M_{CD}=0 \; : \; \frac{M_{AB}+M_{BA}}{10}+\frac{M_{CD}+M_{DC}}{4}=0.$$

Expanding, and substituting for θ_D in order to keep the number of unknown quantities to three, we have

$$+2{\cdot}533\theta_B+0{\cdot}667\theta_C-0{\cdot}180d-9/EI=0.$$

$$+0{\cdot}667\theta_B+2{\cdot}083\theta_C-0{\cdot}188d+9/EI=0.$$

$$+0{\cdot}180\theta_B+0{\cdot}187\theta_C-0{\cdot}083d \qquad =0.$$

Values of Slopes and Sidesway. Solving the three equations simultaneously,

$$\theta_B=+\frac{4{\cdot}95}{EI} \; : \; \theta_C=-\frac{6{\cdot}19}{EI} \; : \; d=-\frac{3{\cdot}26}{EI} \; : \; \theta_D=+\frac{1{\cdot}88}{EI}$$

from which the values of the moments at A, B and C can be found.

Reminder. In all Slope Deflection problems the signs of the various quantities are of importance. The student should be thoroughly familiar with the convention of signs given at the beginning of the chapter, and study all the diagrams with this convention in mind.

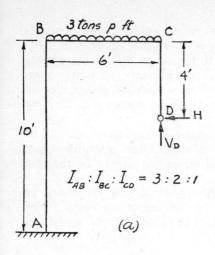

$I_{AB} : I_{BC} : I_{CO} = 3 : 2 : 1$

(a)

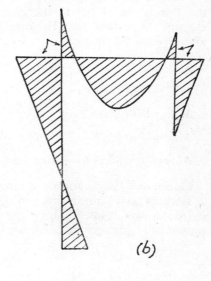

(b)

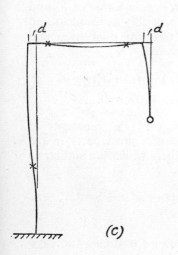

(c)

5.9 This building frame, under a simplified wind load, is symmetrical. Thus the slopes and displacements at B, C and D must be equal, respectively, to the slopes and displacements at G, F and E. It is not, therefore, necessary to write down equations for the whole of the structure, but only for one side of it. The conditions at A, B, C and D determine the bending moments at all sections.

The systematic solution of simultaneous equations should be practised in the form shown in the table. The quantity which appears least often in any group of equations should be the first to be eliminated.

Fundamental Expressions for Moments. From the symmetry of the frame, the slopes at the ends of the beams are equal. For example, θ_G equals θ_B. These relationships simplify the equations and give the following:

$M_{AB}=4EK(2\theta_A+\theta_B-3R_B)$. $M_{CF}=2EK(3\theta_C)$.

$M_{BA}=4EK(2\theta_B+\theta_A-3R_B)$. $M_{CD}=4EK(2\theta_C+\theta_D-3R_D)$.

$M_{BG}=2EK(3\theta_B)$. $M_{DC}=4EK(2\theta_D+\theta_C-3R_D)$.

$M_{BC}=4EK(2\theta_B+\theta_C-3R_C)$. $M_{DE}=2EK(3\theta_D)$.

$M_{CB}=4EK(2\theta_C+\theta_B-3R_C)$.

Conditions of Equilibrium. θ_A might be found in terms of θ_B and R since $M_{AB}=0$, but in this problem it is probably as rapid to find θ_A as shown below. There are seven unknown quantities, and thus seven equations of conditions are required.

1. $M_{AB}=0$.   3. $M_{CB}+M_{CF}+M_{CD}=0$.

2. $M_{BA}+M_{BG}+M_{BC}=0$.   4. $M_{DC}+M_{DE}=0$.

Since $M_{BC}+M_{CB}+M_{GF}+M_{FG}=-1\times10$ and the moments in the right-hand column equal their respective counterparts in the left-hand column, this equation and the others which are similar to it can be written,

5. $M_{BA}+5=0$.   7. $M_{CD}+M_{DC}+5=0$.

6. $M_{BC}+M_{CB}+5=0$.

Values of Slopes and Sideways. Substituting for slopes and sideways in the equations of equilibrium, the seven equations are solved as follows:

116

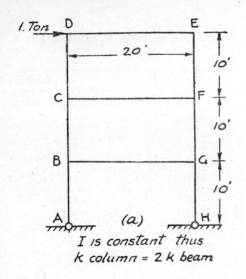

1. Ton D   E

20'

10'

C   F

10'

B   G

10'

A   (a)   H

I is constant thus
k column = 2 k beam

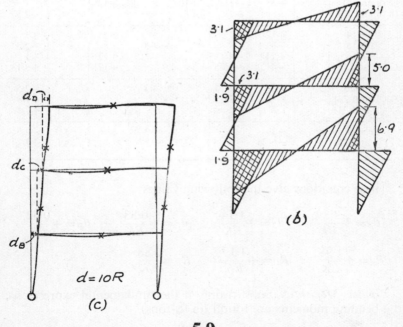

d_D

d_C

d_B

$d = 10R$

(c)

3·1

3·1

3·1

5·0

1·9

6·9

1·9

(b)

Solution of Equations.

Figures represent coefficients of the unknown quantities in equations 1 to 7 above.

	θ_A	θ_B	θ_C	θ_D	R_B	R_C	R_D	Number $\times EK$
1	+2	+1			−3			
2	+4	+22	+4		−12	−12		
3		+4	+22	+4		−12	−12	
4			+4	+14			−12	
5	+4	+8			−12			+5
6		+12	+12			−24		+5
7			+12	+12			−24	+5
8=4			+4	+14			−12	
9=3		+4	+22	+4		−12	−12	
10=7			+12	+12			−24	+5
11=6		+12	+12			−24		+5
12 (1 and 5)		−3			+3			−2·5
13 (2 and 5)		−14	−4			+12		+5
14=11		+12	+12			−24		+5
15=12		−3			+3			−2·5
16=13		−14	−4			+12		+5
17 (8 and 9)		−4	−18	+10		+12		
18 (8 and 10)			−2	+8				−2·5
19=14		+12	+12			−24		+5
20=15		−3			+3			−2·5
21=16		−14	−4			+12		+5
22 (17 and 18)		−4	−15.5			+12		+3·13
23 =20		−3			+3			−2·5
24 (21and 22)		−10	+11·5					+1·87
25 (19 and 21)		−8	+2					+7·5
26=23		−3			+3			−2·5
27 (24 and 25)		−6·26						+7·17

These equations give the following values:

$$\theta_A = +\frac{2\cdot39}{EK} \ : \ \theta_B = +\frac{1\cdot14}{EK} \ : \ \theta_C = +\frac{0\cdot83}{EK} \ : \ \theta_D = +\frac{0\cdot52}{EK}$$

$$R_B = +\frac{1\cdot98}{EK} \ : \ R_C = +\frac{1\cdot19}{EK} \ : \ R_D = \frac{0\cdot88}{EK}.$$

Bending Moments. Substituting in the fundamental expressions, the bending moments are found (in ft.-tons).

5.10 This structure is symmetrical and symmetrically loaded, and thus only one side (ABC) need be considered. $\theta_D = -\theta_C$ and $\theta_A = 0$. The unknown quantities are, therefore, θ_B, θ_C, R_B and R_C. The displacement of B must be horizontally outwards $(10R_B)$, while, from the symmetry of the frame, the displacement of C must be vertically downwards. R_B is thus negative, and R_C is positive. These signs may be incorporated in the equations. If the final solution of these deflections carries a positive sign, the assumptions have been correct.

Value of R_C. From the geometry of the frame outline, it is possible to determine R_C in terms of R_B and thus reduce the number of unknown quantities, and the number of equations required, to three.

In **(b)** the member BC is shown, under the influence of the loading, to be displaced from its original position (BC) to a new position, HM, B moving horizontally and C vertically. Let the horizontal movement of B be d. R_B and R_C will now be found in terms of d.

Consider triangle HRM, where $BR = 5$ ft.; $RC = 5\sqrt{3}$ ft. Then

$$(5+d)^2 + (5\sqrt{3} - CM)^2 = 10^2$$

from which, neglecting small quantities of the second order,

$$CM = \frac{d}{\sqrt{3}} \, ft.$$

for a small movement.

For the Slope Deflection equations the relative movement of C and B, at right angles to CB, is required. The vertical movement of C has a component OM, and the horizontal movement of B a component HL, in the direction required.

The relative displacement of C and B at right angles to CB is

$$OM + HL$$

$$= \frac{1}{2} \cdot \frac{d}{\sqrt{3}} + \frac{\sqrt{3}d}{2} = \frac{2d}{\sqrt{3}} \, ft.$$

Thus R_C, which is this displacement divided by the length of BC, is

$$\frac{d}{5\sqrt{3}} \quad \text{and} \quad 3R_C = \frac{\sqrt{3}d}{5}.$$

119

5.10

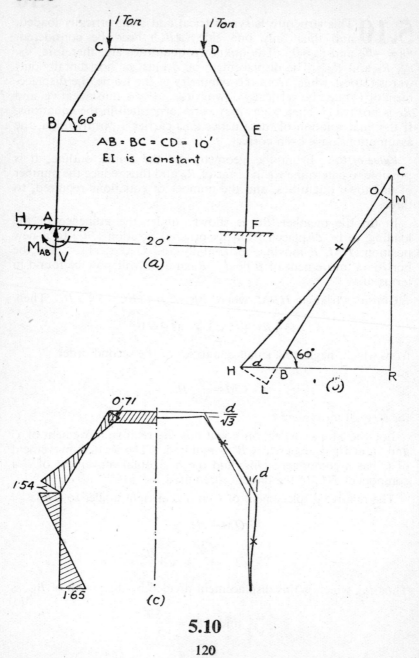

AB = BC = CD = 10'
EI is constant

(a)

(b)

(c)

Fundamental Expressions for Moments.

$$M_{AB}=\frac{EI}{5}\left(\theta_B+\frac{3d}{10}\right) \qquad M_{CB}=\frac{EI}{5}\left(2\theta_C+\theta_B-\frac{\sqrt{3}d}{5}\right)$$

$$M_{BA}=\frac{EI}{5}\left(2\theta_B+\frac{3d}{10}\right) \qquad M_{CD}=\frac{EI}{5}\theta_C$$

$$M_{BC}=\frac{EI}{5}\left(2\theta_B+\theta_C-\frac{\sqrt{3}d}{5}\right).$$

Conditions of Equilibrium. When one or more members of a frame are inclined, it is not sufficient to consider horizontal shear only, for the offsetting of the vertical loading by the inclination of the members has the effect of introducing an extra moment which must be accounted for. Moments must be taken about a sufficient number of points to make it possible to eliminate the unknown forces. Problem **6.14** is a similar type.

Taking moments about *B*

$$M_{AB}-10H=M_{BC}=-M_{BA}.$$

Taking moments about *C*

$$M_{AB}-H(10+5\sqrt{3})+5V=M_{CD}=-M_{CB}.$$

Eliminating *H* from these two equations we have ($V=1$ *ton*)

$$0\cdot465M_{AB}-0\cdot535M_{CB}+M_{BA}-2\cdot68=0.$$

The other two equations of equilibrium are:

$$M_{BA}+M_{BC}=0 \quad : \quad M_{CB}+M_{CD}=0.$$

Values of Slopes and Sidesway. Substituting the fundamental expressions for moments, and simplifying,

$$1\cdot93\theta_B-1\cdot07\theta_C+0\cdot625d=13\cdot40/EI$$
$$4\cdot00\theta_B+1\cdot00\theta_C-0\cdot046d=0$$
$$1\cdot00\theta_B+3\cdot00\theta_C-0\cdot346d=0$$

and from these equations

$$\theta_B=-\frac{0\cdot55}{EI} \quad : \quad \theta_C=+\frac{3\cdot56}{EI} \quad : \quad d=+\frac{29\cdot3}{EI}\ ft.$$

from which the bending moment diagram can be obtained.

FRAMES WITH SETTLEMENT OF FOUNDATIONS

A slight settlement of the foundations of a structure, though having no effect if the frame is statically determinate, makes a considerable difference to the forces and moments induced in the structure if it is continuous, or contains redundant members (Chapter 9).

It is important, then, that the engineer should be able to estimate, from a consideration of the soil mechanics of the foundation strata, what the probable differential settlement will be, and what effect such a settlement will have on the forces and moments in a continuous frame. Problems 5.11 and 5.12 indicate how this effect can be determined.

5.11 In this frame, the hinge E sinks through a distance $\varDelta$, vertically. The ability to draw the probable deflected form of a frame, on which much emphasis has been laid in this book, is of value in a problem of this kind. A sketch of the deformed shape of the frame indicates that A, B and F will each move horizontally through an unknown distance e.

The stiffness of each beam (K) is 4, and that of each column 6 $ft.^3$

Fundamental Expressions for Moments. From the figure we have

$$M_{DA}=12E(2\theta_D+\theta_A-e/4)=0 \qquad M_{CB}=12E(2\theta_C+\theta_B-e/4)=0$$

$$M_{AD}=12E(2\theta_A+\theta_D-e/4) \qquad M_{BF}=8E(2\theta_B+\theta_F-\varDelta/2)$$

$$M_{AB}=8E(2\theta_A+\theta_B) \qquad M_{FB}\doteq8E(2\theta_F+\theta_B-\varDelta/2)$$

$$M_{BA}=8E(2\theta_B+\theta_A) \qquad M_{FE}=12E(2\theta_F+\theta_E-e/4)$$

$$M_{BC}=12E(2\theta_B+\theta_C-e/4) \qquad M_{EF}=12E(2\theta_E+\theta_F-e/4)=0.$$

Conditions of Equilibrium. Each joint is in equilibrium, and the horizontal shear is zero.

From $\qquad M_{DA}=M_{CB}=M_{EF}=0$ we have

$$\theta_D=e/8-\theta_A/2 \ : \ \theta_C=e/8-\theta_B/2 \ : \ \theta_E=e/8-\theta_F/2.$$

Also $M_{AB}+M_{AD}=0 \ : \ M_{BA}+M_{BC}+M_{BF}=0 \ : \ M_{FB}+M_{FE}=0$

$$1/12(M_{DA}+M_{AD}+M_{BC}+M_{CB}+M_{FE}+M_{EF})=0.$$

The unknown quantities are θ_A, θ_B, θ_F and e.

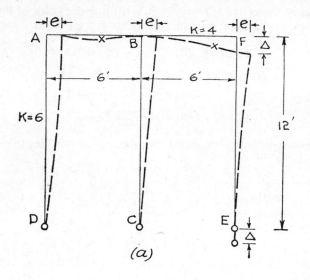

(a)

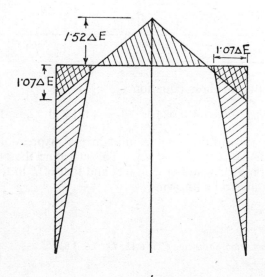

(b)

5.11

5.11

Values of Slopes and Sidesway. Substituting the value of the fundamental expressions in the equations above, we obtain

$$+34\theta_A + 8\theta_B \qquad -1\cdot5e = 0$$
$$+8\theta_A + 50\theta_B + 8\theta_F = 1\cdot5e = 4\varDelta$$
$$+8\theta_B + 34\theta_F - 1\cdot5e = 4\varDelta$$
$$+18\theta_A + 18\theta_B + 18\theta_F - 4\cdot5e = 0.$$

Figures represent coefficients of unknown quantities in equations 1 to 4 above.

No.	θ_A	θ_B	θ_F	e	$\varDelta$	
1	+34	+8	—	−1·5	—	=0
2	+8	+50	+8	−1·5	−4	=0
3	—	+8	+34	−1·5	−4	=0
4	+18	+18	+18	−4·5	—	=0
5 (1 and 2)	+26	−42	−8	—	+4	=0
6 (2 and 3)	+8	+42	−26	—	—	=0
7 (3 and 4)	+18	−6	−84	—	+12	=0
8 (5 and 6)	+34	—	−34	—	+4	=0
9 (6 and 7)	+134	—	−614	—	+84	=0
10	—	—	−480	—	+68	=0

From the solution of these equations,

$$\theta_A = +0\cdot024\varDelta \ : \ \theta_B = +0\cdot084\varDelta \ : \ \theta_F = +0\cdot142\varDelta \ : \ e = \varDelta.$$

From which, by substitution in the fundamental expressions, the bending moments can be determined. To determine the bending moments in ft.-tons, the value of $\varDelta$ in feet and that of E in tons per sq. ft. would require to be known.

Exercises

Solve, by *Slope Deflection*, problems 6.11, 6.12, 7.4, 7.6, 3.8.

5.12 In this frame all the members have equal stiffness (*K*). When loaded, the frame suffers a displacement of the foundations. There is no vertical settlement, but the abutment at *A* rotates clockwise (positive rotation) through $\theta_A = 0 \cdot 15/EK$ radians, and *D* slips horizontally to the right through $0 \cdot 25/EK$ ft. ($\varDelta$). Find the effect on the moments of the frame of this displacement.

When *D* moves towards the right through a known distance $\varDelta$, *B* and *C* must also move towards the right through an unknown distance *e*. The movement $\varDelta$ causes a negative rotation of *CD*, and the movement *e* causes a positive rotation of *AB* and *CD*. The nett relative displacement of *C* and *D* is therefore the algebraic sum of $\varDelta$ and *e*.

Fundamental Expressions for Moments.

$$M_{AB} = 2EK\left(2\theta_A + \theta_B - \frac{3e}{5}\right) \quad : \quad M_{CB} = 2EK(2\theta_C + \theta_B)$$

$$M_{BA} = 2EK\left(2\theta_B + \theta_A - \frac{3e}{5}\right) \quad : \quad M_{CD} = 2EK\left\{2\theta_C + \frac{3(\varDelta - e)}{7}\right\}$$

$$M_{BC} = 2EK(2\theta_B + \theta_C) \quad : \quad M_{DC} = 2EK\left\{\theta_C + \frac{3(\varDelta - e)}{7}\right\}$$

Conditions of Equilibrium.

$$M_{BA} + M_{BC} = 0 \quad : \quad M_{CB} + M_{CD} = 0$$

$$1/5\,(M_{BA} + M_{AB}) + 1/7\,(M_{CD} + M_{DC}) = 0.$$

Values of Slopes and Sidesway. Substituting the known values of θ_A and $\varDelta$ (given at the beginning of the problem) we have

$$+4\theta_B + \theta_C - 0 \cdot 6e + 0 \cdot 15/EK = 0$$

$$+7\theta_B + 28\theta_C - 3e + 0 \cdot 75/EK = 0$$

$$+0 \cdot 6\theta_B + 0 \cdot 429\theta_C - 0 \cdot 362e + 0 \cdot 121/EK = 0.$$

The simultaneous solution of these equations results in

$$\theta_B = +\frac{0 \cdot 016}{EK} \quad : \quad \theta_C = +\frac{0 \cdot 009}{EK} \quad : \quad e = +\frac{0 \cdot 369}{EK}$$

from which the bending moment diagram can be constructed by substitution in the fundamental expressions.

5.12

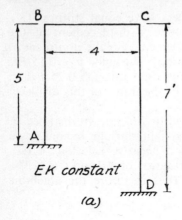

(a)

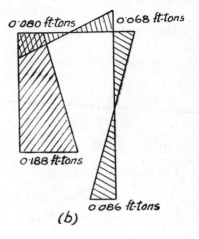

(b)

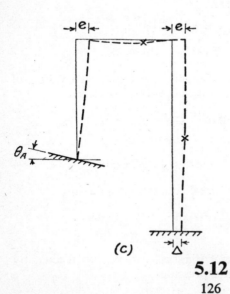

(c)

5.12

Revision Exercises

Solve each of these problems by at least two methods.

A beam AC of $2l$ ft. span carries two loads, each of W tons, at quarter-span points. Show that if an extra support is inserted at the centre of the span (B) so that A, B, and C are at the same level, then the load on the centre support will be $1\cdot375W$. The span is freely supported at A and C.

A beam CA is held horizontally at both ends, and supported to the level of the ends at an intermediate point B. $CA = 15$ ft.; $CB = 7$ ft. The only load is 10 tons concentrated at 4 ft. from C. $\qquad\qquad (M_B = -4\cdot6\ ft.\text{-}tons.)$

A continuous beam $ABCD$ has three spans, each 25 ft. long. The live load on this beam may be 1 ton per ft. run. Draw B.M.D. and S.F.D. for the five cases:

AB loaded; BC loaded; all spans loaded; AB and BC loaded; AB and CD loaded. $\qquad\qquad (M_A = M_D = \text{zero.})$

From these, draw the B.M. envelope diagram whose ordinates indicate the maximum positive and negative moments possible at any section.

	Case	1	2	3	4	5
M_B – – – –		$-41\cdot7$	$-31\cdot3$	$-62\cdot5$	$-72\cdot9$	$-31\cdot3$ ft.-tons
M_C – – – –		$+10\cdot4$	$-31\cdot3$	$-62\cdot5$	$-20\cdot9$	$-31\cdot3$,,

If $AB = CD$ and is half the length of BC (letters refer to Fig. 3.3) and the frame carries a concentrated load W at the centre of BC, show that the horizontal reaction H is three-sixteenths of W.

Draw the B.M.D. and deflected form of the frame when $AB = CD = 6$ ft. and $BC = 5$ ft. The load is 5 tons (concentrated) on BC at 2 ft. from B. $I_{AB} : I_{BC} : I_{CD} = 1 : 2 : 1$. $\qquad\qquad (H = 0\cdot192\ tons.)$

A rigid two-hinged frame has sloping legs AB and CD, and is symmetrical about its centre line. The hinges are on the same level. The height from the hinges to B is 8 ft. 8 in.; $BC = 9$ ft., and the distance between hinges is 19 ft. The beam BC is loaded with two equal vertical loads of 2 tons, at 2 ft. from B and 2 ft. from C respectively. Draw the B.M.D. and deflected form. I is constant throughout. $\qquad\qquad (M_B = 2\cdot27\ ft.\text{-}tons.)$

A horizontal concentrated load of 2 tons is applied to B in a rectangular two-hinged rigid frame. $I_{AB} : I_{BC} : I_{CD} = 1 : 0\cdot5 : 0\cdot8$. $AB = 6$ ft.; $BC = 20$ ft.; $CD = 8$ ft. Draw the B.M.D. and the deflected form of the frame. $\qquad\qquad (M_B = 7\cdot37;\ M_C = 6\cdot18\ ft.\text{-}tons.)$

CHAPTER 6

METHOD OF MOMENT DISTRIBUTION

This method of solving statically indeterminate frames was first suggested and described by Professor Hardy Cross, and is often known as the "Hardy Cross" method. Instead of requiring the solution of simultaneous equations, the Method of Moment Distribution operates by a series of approximations, each of which is successively nearer the exact solution. The determination of the values of these successive approximations depends on two facts which must first be appreciated.

Basic Assumptions

(1) When a stiff joint in a structure absorbs an applied moment with rotational movement only (no translation) the moment is taken up or resisted by the various members of the joint in proportion to their respective stiffnesses. The stiffer the member, the more it resists bending and the greater is the proportion of the applied moment which it can absorb. As in previous chapters, the stiffness is moment of inertia divided by length.

(2) The second fact has been stated and illustrated in previous chapters and should be familiar. When a member is fixed at one end and a moment is applied at the other (free) end in such a way that the free end, while rotating, retains its original position (as if held by a hinge), the moment induced at the fixed end is one-half of the applied moment. Students sometimes find difficulty in visualising the direction or sense of these moments. The following test will demonstrate clearly what occurs:

Hold a flexible scale in a horizontal position by a finger and thumb at each end. Keep the left-hand end horizontal and at the same time apply a clockwise (or positive) moment to the right-hand end. The right hand sinks and the left hand applies a counter-clockwise (or negative) moment in order to keep the left end horizontal. There is no point of contraflexure in the scale.

Now make the member fulfil the condition described above by raising the right hand to its original position, while still applying the positive moment with the right finger and thumb. It will be found that the left hand must immediately change the direction of

128

the induced moment to give a clockwise (or positive) twist in order to keep the left-hand end "fixed," or in a horizontal position. There is now a point of contraflexure at one-third of the length from the left-hand or fixed end.

This result can be summarised by saying that the induced moment at the fixed end is in the same direction as the applied moment and of half its value. The truth of this statement was demonstrated in Problem **2.9** (*BC*) and Problem **3.12** (*AB*).

Convention of Signs

Convention of Signs. The most convenient convention is that used in Chapter 5. A moment is positive if it is applied in a clockwise direction. Fig. **6.01** shows the signs of the "fixing moments" of a loaded beam.

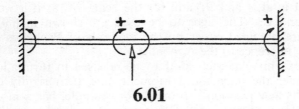

6.01

6.1 Professor Hardy Cross described his method as a physical conception implying that the deformations of the structure under the various conditions imposed must be visualised. The practice obtained in previous chapters in the drawing of deflected forms of beams and frames is of assistance in such visual interpretation.

Fundamental Stages in the Moment Distribution Method

(1) Sketch the beam, reasonably large, and approximately to scale (**6.1**).

(2) Imagine all the three joints A, B and C to be rigidly fixed, with horizontal tangents.

(3) Write down the fixed-end moments for the beam AB as if it were built in at A and B; and for the beam BC as if it were built in at B and C. The signs to be used are shown in **6.01**. The F.E.M. are obtained either by the methods of Problems **2.2** and **3.6**, or by reference to the Appendix.

(4) Next, imagine each joint to be released in turn, allowed to rotate under the unbalanced moment, and then rigidly clamped again in its new position. Joint B, for example, has a moment of -300 ft.-tons on one side of the support and zero on the other side. This unbalanced state of affairs can exist only so long as the joint is held firmly in the assumed position and restrained from rotating.

When joint B is released, the tangent at B rotates clockwise and the two members forming the joint B take up the unbalanced moment (-300 ft.-tons) in proportion to their stiffnesses. In this beam, since the stiffness of BA equals that of BC, the distribution factor for each is one-half.

(5) The step described in (4) is known as "balancing," the balancing moment applied being of the opposite sign to that of the unbalanced moment. At a free end, such as C, the release of the joint results in a zero moment and the balancing moment must be equal to the unbalanced moment.

(6) When all joints have been balanced, *draw a line to indicate the end of this stage*. The drawing of this line is of great assistance in more complex frames and should never be omitted.

(7) During the balancing operation each member of a joint rotates under the influence of the applied balancing moment and thus the second fact considered at the beginning of the chapter comes into play. Moments equal to half of the balancing moments

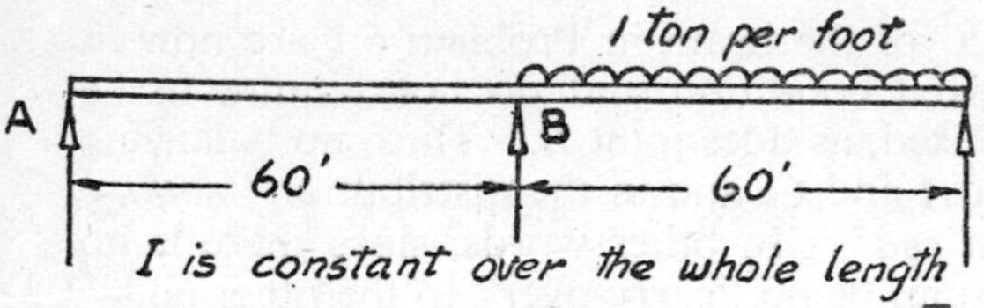

1 ton per foot				
A		B		C
← 60' →		← 60' →		

I is constant over the whole length

—	—	−300·0	+300·0	$\frac{\omega l^2}{12}$ (F.E.M)
—	+150·0	+150·0	−300·0	Balance
+75·0	—	−150·0	+75·0	Distribute
−75·0	+75·0	+75·0	−75·0	Balance
+37·5	−37·5	−37·5	+37·5	Distribute
−37·5	+37·5	+37·5	−37·5	Balance
0	+225	−225	0	Final Bending Moments

6.1

are induced at the adjacent fixed joints as shown by the arrows in **6.1**. This operation is known as "distribution" or "carry over," half the balancing moments being carried over to the opposite end of each member with unchanged sign.

(8) The moments carried over in the previous step appear below the line, representing an unbalanced state which must be eliminated by a further balancing operation. Balancing and distribution are carried out as often as necessary to reduce the residual moment to a negligible amount, the process being stopped after any balance.

(9) Add, algebraically, the moments appearing in the columns at the ends of each member in order to obtain the data necessary for drawing the bending moment diagram.

It has been emphasised in previous chapters that the student should work each problem afresh. This is even more essential in the method of Moment Distribution, where it is impossible to appreciate the work by a mere examination of the diagrams and text.

131

6.2 The ends A and C of the beam in Problem **6.1** are now in a permanently fixed condition and do not require to be alternately released and fixed, as does joint B. Thus, no balancing operation is required for A and C, and in the distribution "*nothing comes back from a fixed end*." In other words, since there is no balancing moment there can be no "carry-over" to the other end.

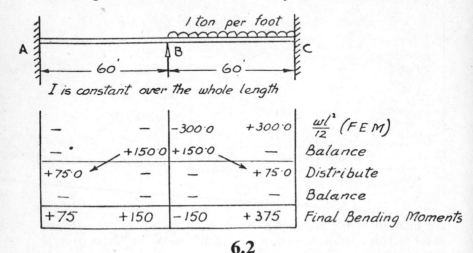

		$-300{\cdot}0$	$+300{\cdot}0$	$\frac{wl^2}{12}$ (F E M)
—	$+150{\cdot}0$	$+150{\cdot}0$	—	Balance
$+75{\cdot}0$	—	—	$+75{\cdot}0$	Distribute
—	—	—	—	Balance
$+75$	$+150$	-150	$+375$	Final Bending Moments

6.2

Simplified Procedure for Beams with One End freely supported

When joints such as B in Problem **6.1** are freed in the "balancing" operation, all the other joints are considered to be held fixed. Thus, the balancing moment applied to BA or BC is applied to one end of a member whose other end is built in. Such a moment M applied at X (Fig. **6.02**) results in a bending moment diagram and deflected form of the type shown at (**a**).

The slope of the beam at X is $\phi_B = \dfrac{1}{EI}\left(\dfrac{3}{2}\dfrac{Ml}{2} - \dfrac{Ml}{2}\right) = \dfrac{Ml}{4EI}$.

Moment Distribution operations, however, would be made much simpler if points such as A and C (**6.1**) could be left permanently free throughout the calculation instead of being alternately fixed and freed as in Problem **6.1**. Fig. **6.02b** shows the effect of freeing the end Y. The slope ϕ_F can be written as d/l where d is represented by the moment of the M/EI diagram about Y.

The slope of the beam at X is $\phi_F = \dfrac{1}{lEI}\left(\dfrac{Ml}{2} \cdot \dfrac{2l}{3}\right) = \dfrac{Ml}{3EI}$, or 4/3 of the rotation of the end X when Y was fixed.

132

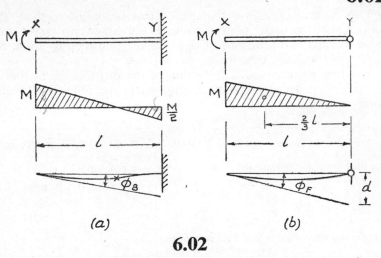

(a)   *(b)*

6.02

Again, when a beam in a Moment Distribution determination carries a load, the first step in the work is to imagine the beam held horizontally at the ends by the "fixed-end" moments which apply to such a loading on a built-in beam. If one end is to be considered permanently free, there can be a fixing moment at the other end only. The magnitude of this fixing moment is increased from its normal "fixed-end" value by half the value of the moment released at the other end. The truth of this can be realised by a study of Problems **2.1** and **3.1**.

In Problem **2.1**, if both ends of the beam were fixed in position and direction, the fixed-end moments would be $wl^2/12$ ($=100/6$) at both ends. When the end B is released to the free condition, the moment at A increases to $\dfrac{100}{6} + \dfrac{1}{2} \cdot \dfrac{100}{6} = 25$ *ft.-tons*.

Summarising these two points, we have the following conclusions. If instead of alternately fixing and freeing the end of a member it is left permanently free throughout the whole of the moment distribution determination, two changes must be made.

(1) *In determining the distribution factors for use in the balancing operation, the stiffness of the member must be reduced to three-quarters of its original value.*

(2) *When the "fixed-end moments" are written down, half of the moment released at the permanently free end must be added to the fixed end moment at the fixed end.*

133

6.3 As an example of the use of this simplified method, the problem under consideration is solved in two ways; first as in the first two problems of this chapter, and then by using the shorter method just described.

The first requirement is the value of the distribution factors at the joint B. These are best determined in the following tabular form:

Joint	Member	Relative Stiffness	Sum	Distribution Factors

For the normal method of Moment Distribution as in Problem **6.1**:

Joint	Member	Relative Stiffness	Sum	Distribution Factors
B	BA	$\dfrac{I}{15}=\dfrac{2}{30}I$	$\dfrac{7}{30}$	$\dfrac{2}{7}=0\cdot286$
	BC	$\dfrac{2I}{12}=\dfrac{5}{30}I$		$\dfrac{5}{7}=0\cdot714$

For the shorter method, when A is permanently free:

Joint	Member	Relative Stiffness	Sum	Distribution Factors
B	BA	$\dfrac{3}{4}\times\dfrac{I}{15}=\dfrac{3I}{60}$	$\dfrac{13}{60}$	$\dfrac{3}{13}=0\cdot23$
	BC	$\dfrac{2I}{12}=\dfrac{10I}{60}$		$\dfrac{10}{13}=0\cdot77$

The balancing and distribution steps follow the same procedure as in Problem **6.1**. For example, in (a) there is an unbalanced moment at B of $-24+15=-9$ *ft.-tons*. The balancing moments must therefore both carry a positive sign. $0\cdot714\times9=6\cdot4$ *ft.-tons* and $0\cdot286=9=2\cdot6$ *ft.-tons*.

In Fig. (b), since the end A has been permanently freed no operation need be carried out there, as no moment is carried over to a free end.

Exercises

Solve, by *Moment Distribution*, problems 2.1, 2.5, 2.6, 2.10, 3.2.

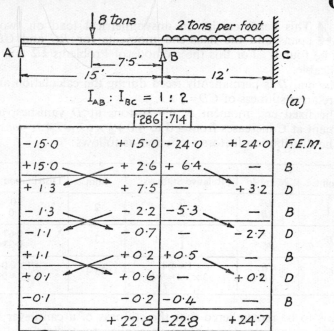

$$I_{AB} : I_{BC} = 1 : 2$$

(a)

·286	·714			
−15·0	+ 15·0	−24·0	+ 24·0	F.E.M.
+15·0	+ 2·6	+ 6·4	—	B
+ 1·3	+ 7·5	—	+ 3·2	D
− 1·3	− 2·2	−5·3	—	B
−1·1	− 0·7	—	− 2·7	D
+1·1	+ 0·2	+0·5	—	B
+ 0·1	+ 0·6	—	+0·2	D
−0·1	− 0·2	−0·4	—	B
O	+ 22·8	−22·8	+24·7	

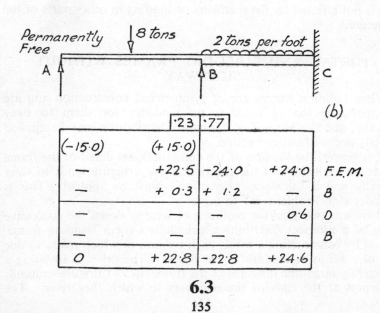

(b)

·23	·77			
(−15·0)	(+ 15·0)			
—	+22·5	−24·0	+ 24·0	F.E.M.
—	+ 0·3	+ 1·2	—	B
—	—	—	0·6	D
—	—	—	—	B
O	+ 22·8	−22·8	+24·6	

6.4 This beam carries an unsymmetrical load on two of its spans and the fixed-end moments for BC and CD must first be found. For this the methods of Problems **2.2** and **3.6** are applicable.

The end D is permanently freed during the calculation and thus the relative stiffness of $CD = \frac{3}{4} \times 3$.

The fixed end moment of 6·8 ft.-tons at D vanishes, and the moment at C increases from 4·0 to $4·0 + \frac{1}{2} \times 6·8 = 7·4$ *ft.-tons*.

The distribution factors are found as follows:

Joint	Member	Relative Stiffness	Sum	Distribution Factors
A	AE AB	0 2	2	0 1·0
B	BA BC	2 1	3	0·67 0·33
C	CB CD	1=4/4 3×3/4=9/4	13/4	4/13 =0·31 9/13 =0·69

It is to be remembered that the stiffness of a cantilever is zero, since it cannot resist any moment applied to its fixed end. Also, the moment at the fixed end of a cantilever is statically determinate and is not affected by the members or loading in other parts of the structure.

PORTAL AND BUILDING FRAMES WITHOUT SIDESWAY

When building frames are of symmetrical construction and are symmetrically loaded there is no tendency for them to sway laterally, and the Method of Moment Distribution may be applied directly and without correction.

If, however, the loading or the constructional detail of the frame is unsymmetrical about the centre line the structure tends to sway laterally and a "sidesway correction" must be applied. This is considered in Problems **6.7** to **6.14**.

There are two possible methods of writing down the successive steps of a moment distribution calculation for a building frame. One is by constructing a table, each column of which refers to one end of a beam or column of the frame. The other is by using a sufficiently large line diagram of the frame and writing the columns of figures at the ends of the members to which they refer. The

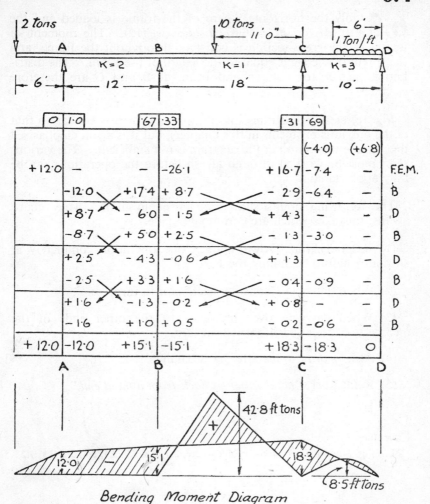

Bending Moment Diagram

latter method has the disadvantage of sometimes requiring a large diagram, but it is more readily comprehended and is more direct than the other method, and will be used in this chapter.

The columns of figures must be written in such a way that they do not foul each other. Usually three or four "balances" are all that is required. The directions in which the columns are written in this chapter are obtained by placing thumb and forefinger at the ends of a beam or column and turning the hand clockwise.

6.5 Only the horizontal beam of the frame is loaded and the F.E.M. for $BC=16\cdot67$ $ft.$-$tons$ $(wl^2/12)$. The moment of inertia of the cross-section is constant throughout the frame, and the length of the leg is equal to the length of the beam. The distribution factors for column and beam at B and C are therefore each $0\cdot5$.

Fig. (c) clearly illustrates the procedure (which is similar to that of the previous problems in this chapter), but it is again emphasised that a mere inspection of the diagram is not sufficient. The various steps must be worked through afresh before the operations can be fully appreciated.

The following points should be borne in mind. They guard against mistakes which are commonly made.

(1) *Be systematic.* Always finish one operation completely at all points before beginning the next.

(2) When balancing, consider each *joint* in turn.

(3) When carrying over moments to the other ends of the members, consider each *member* in turn.

(4) Always *draw a line* after each balance.

(5) Remember, "*nothing comes back from a fixed end.*"

Exercises

Solve, by *Moment Distribution*, problems 2.2, 2.7, 2.8, 2.9, 3.1, 3.6, 5.1, 7.1, 7.2.

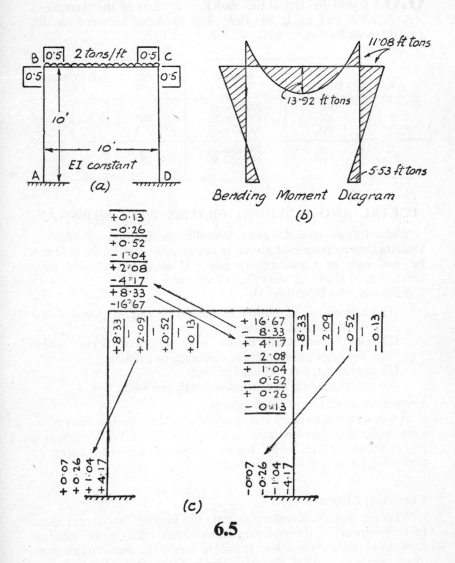

(a)

Bending Moment Diagram

(b)

(c)

6.5

139

6.6 In this frame the moment of inertia of each of the beams *CD* and *BE* is 480 in.[4] and that of each of the stanchions *AB*, *BC*, *DE* and *EF* is 300 in.[4] The fixed-end moment for *BE* $(wl^2/12) = 96$ *ft.-tons.*

Joint	Member	Relative Stiffness	Sum	Distribution Factors
B	*BA*	300/15=20		1/3
	BE	480/24=20	60	1/3
	BC	300/15=20		1/3
C	*CB*	300/15=20		1/2
	CD	480/24=20	40	1/2

PORTAL AND BUILDING FRAMES WITH SIDESWAY

When frames tend to sway laterally the method of Moment Distribution as described above is not directly applicable, and can be used only as a preliminary step. It must be followed by a correction to allow for the effect of sidesway.

Sidesway may be caused by

(1) unsymmetrical outline of the frame (e.g. a portal frame with unequal columns);

(2) non-uniform construction (e.g. a portal frame whose columns have different moments of inertia);

(3) unsymmetrical vertical loading;

(4) lateral loading (e.g. wind or earth pressure),

or by a combination of these causes.

The first step is to carry out the Moment Distribution determination in the way demonstrated in Problems **6.5** and **6.6**, neglecting any sidesway effect. The figures obtained are then corrected according to the following method.

Correcting Moments

Whether there is sidesway or not, the internal horizontal shear in each storey of the building frame must balance the external horizontal shear, since the frame is finally in equilibrium even after sidesway. After the preliminary Moment distribution on a frame subject to sidesway, the apparently unbalanced shear must be corrected by a second distribution. Students should study Fig. **5.0** and the note preceding **5.4**.

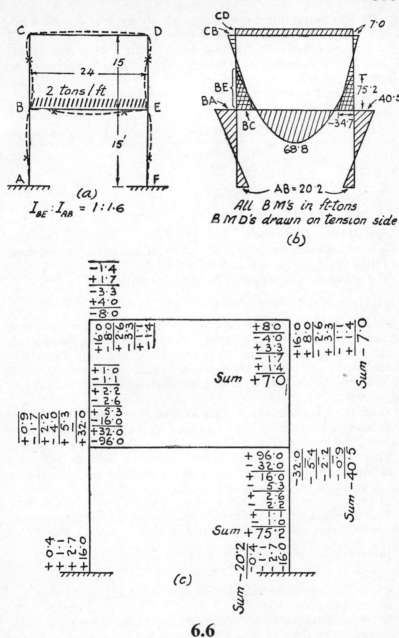

(a)

$I_{BE} : I_{AB} = 1 : 1.6$

(b)

All B M's in ft-tons
B M D's drawn on tension side

(c)

6.7 This portal frame is unsymmetrically loaded, and the fixed-end moments are found by the methods of Problems **2.2** or **3.6** or by reference to the Appendix. The F.E.M. are 3·60 at B and 2·40 ft.-tons at C. The appropriate signs are inserted as shown in Fig. **6.7**.

Distribution Factors (short method of Problem **6.3** used).

Joint	Member	Relative Stiffness	Sum	Distribution Factors
B	BA	$\frac{I}{6} \times \frac{3}{4} = \frac{5}{40} I$	$\frac{13}{40}$	$\frac{5}{13} = 0.385$
	BC	$\frac{I}{5} = \frac{8}{40} I$		$\frac{8}{13} = 0.615$

First Moment Distribution. The preliminary Moment Distribution is carried out, neglecting the effect of sidesway, as was done in Problem **6.5**. This is shown at (**d**).

The unbalanced shear given by this operation (see Fig. **5·0**) is

$$\tfrac{1}{6}(M_{AB}+M_{BA}+M_{CD}+M_{DC})=\tfrac{1}{6}(+1\cdot83-1\cdot47)=+0\cdot06 \ tons.$$

Horizontal Shear. Since no horizontal force acts on the frame (apart from the two forces H, which are in equilibrium) the resultant horizontal internal shear in the frame should be zero. The above determination shows that there is an unbalanced horizontal shear of 0·06 tons towards the right, when the effect of sidesway is neglected.

Since there is no external force of 0·06 tons acting towards the left, correcting moments must be added to the tops of the columns to bring the internal shear to zero.

The moments of inertia and the lengths of the legs are equal and thus, from symmetry, the correcting moment on BA must be equal to the correcting moment on CD.

The total correcting moment $= -0\cdot06$ tons $\times 6$ ft. $= -0\cdot36$ *ft.-tons.*

Therefore, the final moments are obtained by adding algebraically to the moments shown in (**d**) a value of 0·18 ft.-tons at B and C.

Final moment at $B = +1\cdot83 - 0\cdot18 = +1\cdot65$ *ft.-tons.*

    "   " $C = -1\cdot47 - 0\cdot18 = -1\cdot65$ *ft.-tons.*

These two values should be numerically equal, since they are both equal to $6H$ ft.-tons.

(**c**) shows how the frame sways laterally towards the right under the influence of the eccentric load.

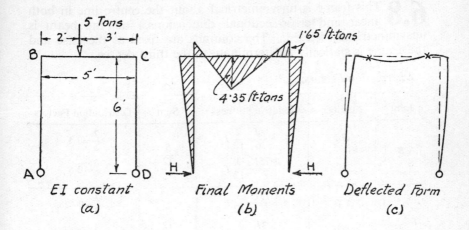

EI constant	*Final Moments*
(a)	*(b)*

Deflected Form
(c)

Sum −1·83

+0·04
−0·07
+0·21
−0·34
+0·46
−0·74
+2·21
−3·60

+1·39 +0·28 −0·13 +0·03

Sum +1·83

+2·40
−1·48
+1·10
−0·68
+0·23
−0·14
+0·11
−0·07

Sum +1·47

−0·92 −0·42 −0·09 −0·04

Sum −1·47

Zero Moment   Zero Moment

(d)

6.7

143

6.8 This frame is unsymmetrical about the centre line in both linear and cross-sectional dimensions, and the beam is unsymmetrically loaded. The columns are fixed at the base and the frame is statically indeterminate to the third degree.

Distribution Factors.

Joint	Member	Relative Stiffness	Sum	Distribution Factors
B	BA	$\dfrac{I}{6}=\dfrac{2}{12}I$	$\dfrac{5}{12}$	$\dfrac{2}{5}=0{\cdot}4$
	BC	$\dfrac{1{\cdot}75I}{7}=\dfrac{3}{12}I$		$\dfrac{3}{5}=0{\cdot}6$
C	CB	$\dfrac{1{\cdot}75I}{7}=\dfrac{3}{12}I$	$\dfrac{7}{12}$	$\dfrac{3}{7}=0{\cdot}428$
	CD	$\dfrac{3I}{9}=\dfrac{4}{12}I$		$\dfrac{4}{7}=0{\cdot}572$

The fixed-end moments for the beam are again found by the methods of Problems **2.2** or **3.6**, or by reference to the Appendix. The F.E.M's. are $-3{\cdot}48$ ft.-tons at B and $+6{\cdot}42$ tons at C.

First Moment Distribution. With these figures as a basis the preliminary moment distribution is carried out, neglecting these effect of sidesway (Fig. (**b**)). The resulting unbalanced shear is

$$\frac{1{\cdot}04+2{\cdot}07}{6}+\left(\frac{-4{\cdot}55-2{\cdot}28}{9}\right)=-0{\cdot}24 \; tons.$$

This must be reduced to zero by correcting moments.

Distribution of Correcting Moments. In Problem **6.7** the columns of the frame were equal in length, and the correcting moment was equally applied to both columns. In this problem the lengths of the columns are unequal and the correcting moment must be divided between the columns in suitable proportions.

In (**c**) positive moments are applied to AB and CD so that the length BC remains unaltered, the displacement of B being equal to that of C.

By Area Moments (Chapter 2), referring to (**d**), the displacement of E from the tangent at B (**c**) is given by the moment of the M/EI diagram between B and E about E.

$$\frac{\Delta}{2}=\frac{M_S}{EI_S}\cdot\frac{S}{4}\cdot\frac{S}{3}=\frac{M_S S^2}{12EI_S}.$$

144

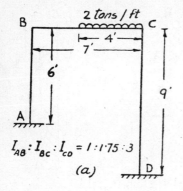

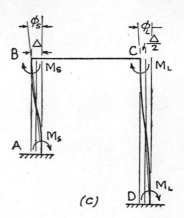

(a)

$I_{AB} : I_{BC} : I_{CD} = 1 : 1·75 : 3$

(c)

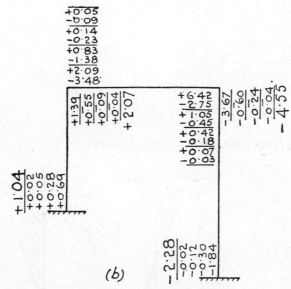

(b)

First Moment Distribution

6.8

6.8

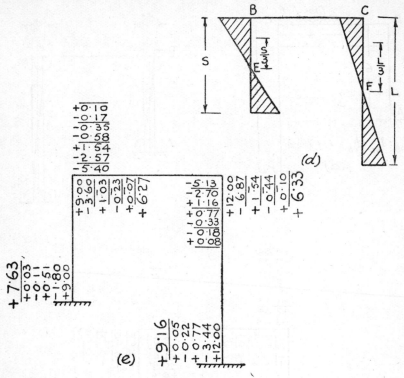

(d)

Correcting Moment Distribution

(e)

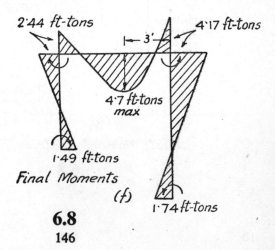

2·44 ft-tons

4·17 ft-tons

3'

4·7 ft-tons max

1·49 ft-tons

1·74 ft-tons

Final Moments

(f)

6.8

Similarly
$$\frac{\Delta}{2} = \frac{M_L L^2}{12 E I_L}$$

or
$$\frac{M_S}{M_L} = \frac{L^2 I_S}{I_L S^2} = \frac{81 \times 1}{3 \times 36} = \frac{3}{4}.$$

Thus, for a frame of this type, the applied correcting moments must vary as

I/length^2 or as stiffness/length.

Another method of obtaining the ratio of the moments to be applied to the two columns is by considering the angles through which AB and CD turn, while suffering a relative displacement of the ends of Δ.

By the method of Slope Deflection (Chapter 5)

$$M_S = 2EK_S(2\theta_A + \theta_B + 3\Delta/S)$$

but
$$\theta_A = \theta_B = 0.$$
$$M_S = 6EK_S \, \Delta/S.$$

Δ/S also represents the final slope of the line joining A and B which, being small, can be written in polar notation, say ϕ_S.

Thus
$$M_S = 6EK_S\phi_S.$$

Similarly
$$M_L = 6EK_L\phi_L$$

and
$$\frac{M_S}{M_L} = \frac{K_S\phi_S}{K_L\phi_L} \text{ or } = \frac{K_S L}{K_L S},$$

i.e. the moments vary as stiffness/length.

Values of Correcting Moments.

Fig. (e) shows how any arbitrary moments, in the ratio of 3 : 4 in this instance, are applied and distributed in the usual way. The figures 9 ft.-tons and 12 ft.-tons were considered to be convenient, but 27 and 36 or 18 and 24 ft.-tons might equally well have been used.

The unbalanced shear produced by these correcting moments is

$$\frac{6 \cdot 27 + 7 \cdot 63}{6} + \frac{9 \cdot 16 + 6 \cdot 33}{9} = +4 \cdot 04 \text{ tons.}$$

The required correcting shear, however, is $+0 \cdot 24$ tons. Thus, if the final moments shown in (e) can produce a correcting shear of $+4 \cdot 04$ tons, by simple proportion it is possible to determine the moments which will produce a correcting shear of $0 \cdot 24$ tons.

6.9

The correcting moments which must be added algebraically to the moments obtained in the first Moment Distribution are shown in the following table, together with the final result:

	First M.D. Result (1)	Arbitrary Corr. Moments (2)	Col. (2) ×0·24/4·04 (3)	Final Result: Sum of (1) and (3)
M_{AB}	+1·04	+7·63	+0·45	+1·49
M_{BA}	+2·07	+6·27	+0·37	+2·44
M_{CD}	−4·55	+6·33	+0·38	−4·17
M_{DC}	−2·28	+9·16	+0·54	−1·74

Sidesway is towards the left. Draw the deflected shape of the frame, marking points of contraflexure.

6.9 Numerous two-span rigid frame bridges have been built. This problem gives the solution of a simplified frame of this type under eccentric loading.

Distribution Factors.

Joint	Member	Relative Stiffness	Sum	Distribution Factors
B	BA	$I/9 \times 2 = 0.055I$	0·138	0·4
	BC	$I/12 = 0.083I$		0·6
C	CB	$I/12 = 0.083I$		0·375
	CD	$I/9 \times 2 = 0.055I$	0·221	0·250
	CE	$I/12 = 0.083I$		0·375

First Moment Distribution. In (c), commencing with F.E.M.'s of −12 and +12 at the ends of *CE*, the preliminary moment is made, neglecting the effect of sidesway. The moments in the three columns give an unbalanced horizontal shear of

$$+\frac{1}{9}(-1·27-0·64+4·34+2·17-6·07-3·04) = -\frac{4·51}{9} \ tons.$$

148

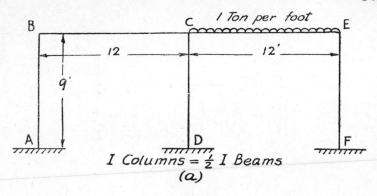

I Columns = ½ I Beams

(a)

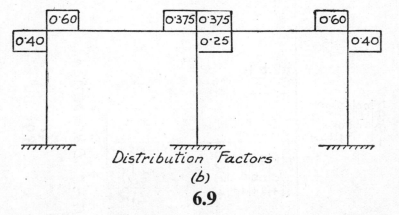

Distribution Factors

(b)

6.9

Values of Correcting Moments. A positive shear of this amount must now be applied by correcting moments. Since all three columns are of equal lengths and moments of inertia, each will take one-third of the required moment (applied equally to top and bottom).

Using an arbitrary figure of 10 ft.-tons, the second Moment Distribution is effected **(d)**.

The moments in the columns now give a correcting shear of

$$\frac{1}{9}(6\cdot38+8\cdot20+8\cdot87+9\cdot43+6\cdot38+8\cdot20)=+\frac{47\cdot46}{9}\ tons.$$

The required correcting shear is 4·51/9 tons, and the moments of (d) must be multiplied by 4·51/47·46 and added algebraically to the moments found in the first distribution.

149

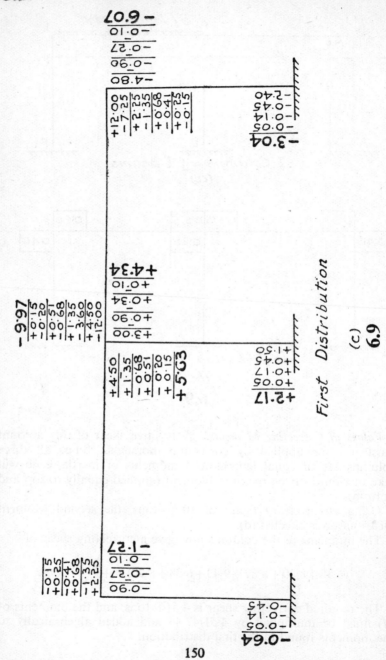

6.9

First Distribution

(c)
6.9

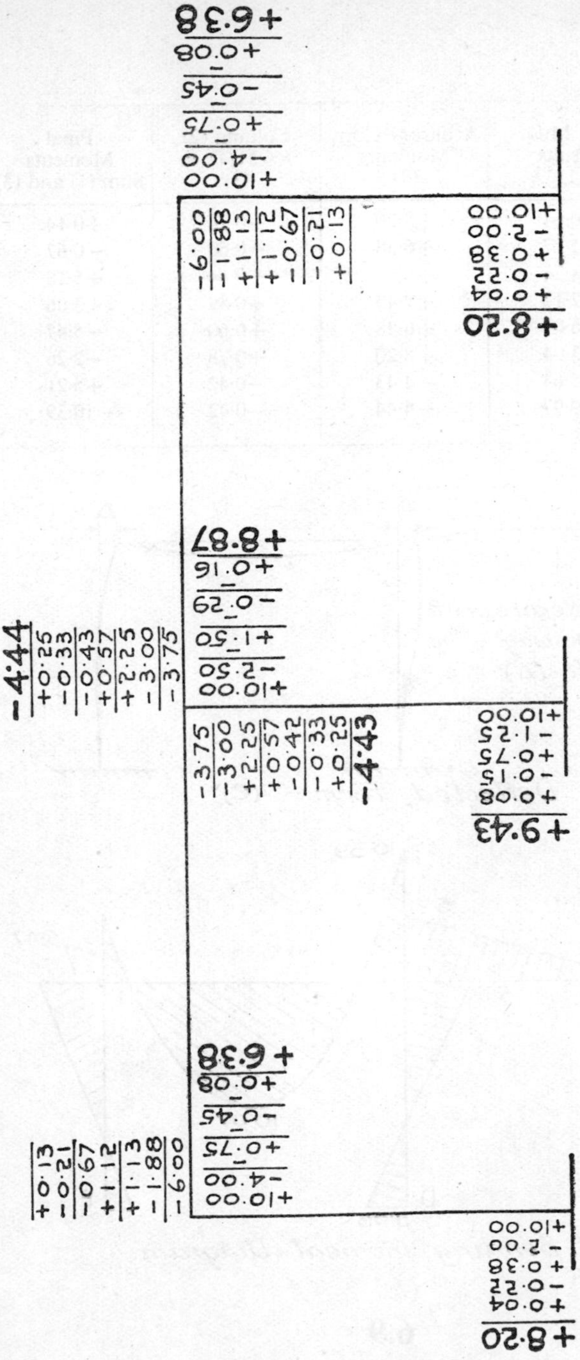

Correction Moment Distribution

(d)
6.9

+6·38
+0·08
−0·45
+0·75
−4·00
+10·00

−6·00
−1·88
+1·13
+1·12
−0·67
−0·21
+0·13

+8·20
+10·00
−2·00
+0·38
−0·22
+0·04

+8·87
+0·16
−0·29
+1·50
−2·50
+10·00

−4·44
+0·25
−0·33
−0·43
+0·57
+2·25
−3·00
−3·75

−4·43
−3·75
−3·00
+2·25
+0·57
+0·42
−0·33
+0·25

+9·43
+10·00
−1·25
+0·75
−0·15
+0·08

+6·38
+0·08
−0·45
+0·75
−4·00
+10·00

+0·13
−0·21
−0·67
+1·12
+1·13
−1·88
−6·00

+8·20
+10·00
−2·00
+0·38
−0·22
+0·04

	First M.D. Results (1)	Arbitrary Corr. Moments (2)	Column (2) ×4·51/47·46 (3)	Final Moments Sum (1) and (3)
M_{AB}	−0·64	+8·20	+0·78	+0·14
M_{BA}	−1·27	+6·38	+0·60	−0·67
M_{CD}	+4·34	+8·87	+0·84	+5·18
M_{DC}	+2·17	+9·43	+0·89	+3·06
M_{EF}	−6·07	+6·38	+0·60	−5·47
M_{FE}	−3·04	+8·20	+0·78	−2·26
M_{CB}	+5·63	−4·43	−0·42	+5·21
M_{CE}	−9·97	−4·44	−0·42	−10·39

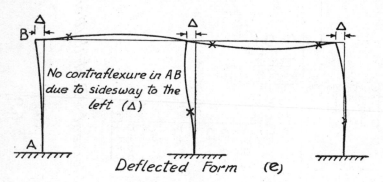

No contraflexure in AB due to sidesway to the left (Δ)

Deflected Form (e)

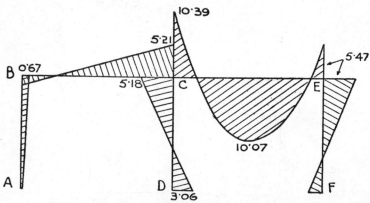

Final Bending Moment Diagram
(f)

6.10 When a building frame consists of more than one storey and is loaded with vertical eccentric loads, the shears in each storey must balance independently of the other storeys. After the first Moment Distribution there will be a residual unbalanced shear in all storeys.

The second step—of applying correction moments to the stanchions—must be made in stages, each storey being treated separately. The final correcting moments are found by multiplying the results obtained by a different factor for each storey.

First Moment Distribution. The F.E.M.'s at the ends of the loaded beam are each $Wl/8 = 10.5$ *in.-lb.* The first distribution is done in the usual way, and gives the following unbalanced shears:

Upper storey:

$$\tfrac{1}{6}(+0.63+4.13-0.74-3.46-0.13+0.33) = +0.127 \ lb.$$

Lower storey:

$$\tfrac{1}{6}(+4.97+2.48-4.03-2.02+0.47+0.24) = +0.352 \ lb.$$

Correcting Moments. Arbitrary correcting moments are now applied to each storey in turn. Since all stanchions are of the same length and moment of inertia the moments applied will be the same for each and will appear in equal amounts, top and bottom, both ends being restrained.

This step requires two complete and distinct Moment Distributions, one for each storey, as is shown in (c) and (d).

The following unbalanced shears are found:

Correction moments applied to upper storey (c):

Unbalanced shear in upper storey :

$$\tfrac{1}{6}(-4.63-3.41-6.01-5.76-4.63-3.41) = -4.642 \ lb.$$

Unbalanced shear in lower storey:

$$\tfrac{1}{6}(+1.33+2.64+1.15+2.29+1.33+2.64) = +1.897 \ lb.$$

Correction moments applied to lower storey (d):

Unbalanced shear in upper storey:

$$\tfrac{1}{6}(+3.40+0.72+2.56+0.98+3.40+0.72) = +1.963 \ lb.$$

Unbalanced shear in lower storey:

$$\tfrac{1}{6}(-8.03-6.07-8.63-7.28-8.03-6.07) = -7.352 \ lb.$$

The required correction shear is -0.127 lb. in the upper storey and -0.352 lb. in the lower storey. The figures found in (c) must thus be multiplied by one factor and those found in (d) by another factor, whose values are such that the required correcting shear is obtained.

6.10

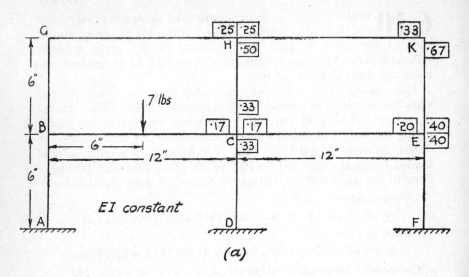

(a)

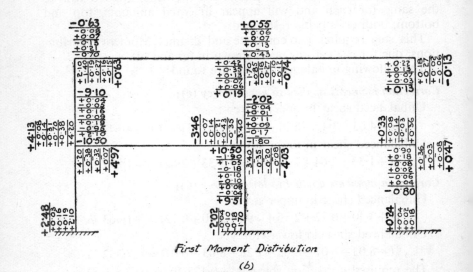

First Moment Distribution

(b)

6.10

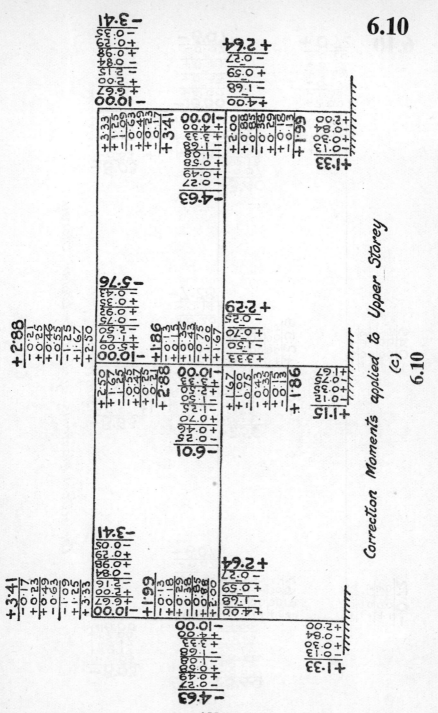

Correction Moments applied to Upper Storey

(c)

6.10

6.10

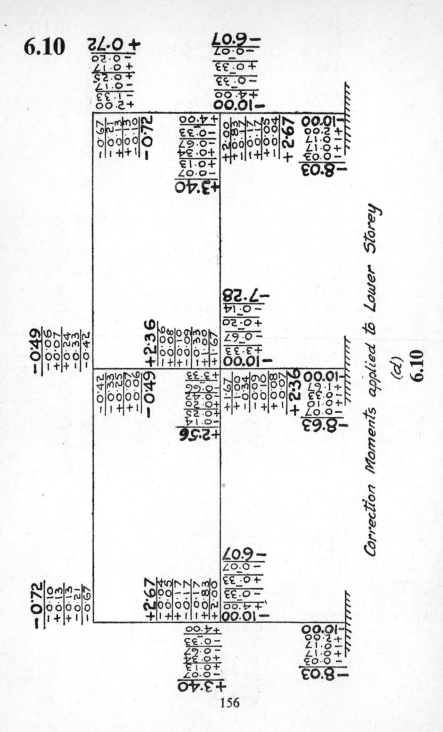

Correction Moments applied to Lower Storey

(d)

6.10

Values of Factors Required.

Let x be the factor by which the moments of (c) are multiplied.

„ y „ „ „ „ (d) „

The results required are, for the upper storey,

$$-4.642x + 1.963y = -0.127,$$

and for the lower storey,

$$+1.897x - 7.352y = -0.352.$$

Solving these equations simultaneously we have,

$$x = 0.053 \qquad\qquad y = 0.062.$$

Multiplying the figures of (c) by this value of x, and the figures of (d) by the value found for y, the correcting moments are obtained. These correcting moments must then be added algebraically to the moments found in the first Moment Distribution.

Moment at	First Moment Distribution	Correcting Moments from × by 0·053	Correcting Moments from × by 0·062	Final Moments: Sum of first three columns in.-lb.
AB	+2·5	+0·1	−0·5	+2·1
DC	+2·0	+0·0	−0·5	2·5
FE	+0·2	+0·1	−0·5	−0·2
BA	+5·0	+0·1	−0·4	+4·7
BG	+4·1	−0·2	+0·2	+4·1
BC	−9·1	+0·1	+0·2	−8·8
CH	−3·5	−0·3	+·02	−3·6
CE	−2·0	+0·1	+0·1	−1·8
CD	−4·0	+0·1	−0·5	−4·4
CB	+9·5	+0·1	+0·2	+9·8
EK	+0·3	−0·2	+0·2	+0·3
EF	+0·5	+0·1	−0·4	+0·2
EC	−0·8	+0·1	+0·2	−0·5
GB	+0·6	−0·2	+0·1	+0·5
GH	−0·6	+0·2	−0·1	−0·5
HG	+0·2	+0·1	−0·0	+0·3
HK	+0·6	+0·1	−0·0	+0·7
HC	−0·7	−0·3	+0·0	−1·0
KH	+0·1	+0·2	−0·0	+0·3
KE	−0·1	−0·2	+0·0	−0·3

6.10

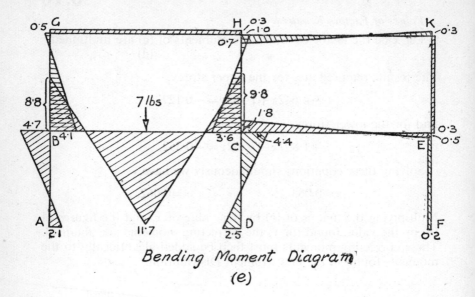

Bending Moment Diagram

(e)

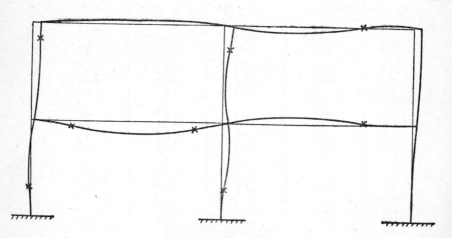

Deflected Form

(f)

6.10

Exercises

Solve, by *Moment Distribution*, problems 3.3, 3.4, 3.5, 3.7, 5.8, 7.3, 7.4, 7.6.

PORTAL AND BUILDING FRAMES WITH LATERAL LOADING

On the frames previously discussed in this chapter only vertical loads have been used, and in the absence of a horizontal component, the horizontal shears in the frames have been zero. In the presence of a horizontal or inclined loading, however, the horizontal shear brought into play by the moments developed must balance the external horizontal shear caused by the loading.

6.11 *The rectangular rigid reinforced concrete frames of a small bridge are spaced 7 ft. 6 in. apart. The bridge carries the Ministry of Transport uniformly distributed and concentrated loads and, in addition, is subjected to earth pressure due to a filling weighing 120 lb. per cu. ft. ($\phi=30°$). Determine whether the size of the beam and the main reinforcement are adequate when it is assumed that the filling is not in contact with the side CD and the bridge is subjected to earth pressure from the left only.*

Fixed-end Moments for BC and AB.

Beam *BC*:

Dead load (road material, slab and rib), say, 1250 *lb./ft.*

Equiv. U.D.L. (Ministry of Transport) $\qquad$ 1650 „

$\qquad\qquad\qquad\qquad\qquad\qquad\qquad\qquad$ ――――

$\qquad\qquad\qquad\qquad\qquad\qquad\qquad\qquad$ 2900 „

$\qquad\qquad\qquad\qquad\qquad\qquad$ or 1·3 *tons per ft.*

Knife edge load (M.o.T.)$=7\cdot5\times2700=9$ *tons.*

$$\text{Fixed-end moment}=\frac{wl^2}{12}+\frac{Wl}{8}=\frac{1\cdot3\times15^2}{12}+\frac{9\times15}{8}=41\cdot30 \, ft\text{.-}tons.$$

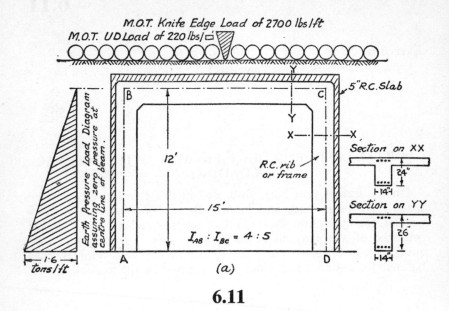

6.11

Side *AB*:

According to Rankine, earth pressure at a depth of 12 *ft.*

$$=120 \times 12 \times \frac{1-\sin 30°}{1+\sin 30°} = 480 \ lb. \ per \ sq. \ ft.$$

The frames are 7·5 ft. apart, and the intensity of horizontal thrust at *A* is, therefore, 1·6 tons per ft. of height. The loading diagram is as shown in (**a**), assuming that the earth pressure is zero at the centre line of *BC*.

From the Appendix or by Area Moments the F.E.M.'s are:

$$M_A = \frac{Wl}{10} = \frac{1·6 \times 12}{2} \times \frac{12}{10} = 11·50 \ ft.\text{-}tons.$$

$$M_B = Wl/15 \qquad \qquad = 7·67 \quad ,,$$

Distribution Factors. Moment of inertia for a rectangular section varies as the cube of the depth, breadth being constant.

Thus $I_{AB} : I_{BC} = 24^3 : 26^3 = 4 : 5.$

Also $l_{AB} : l_{BC} = 4 : 5$ and distribution factors are thus 0·5.

First Moment Distribution. The first unbalanced moment at B is $-41\cdot30+7\cdot67$ ft.-tons, and the balancing moments are each equal to one-half of this. At C the unbalanced moment is $41\cdot30$ ft.-tons.

The first Moment Distribution gives an unbalanced shear of (from **(b)**)

$$\tfrac{1}{12}(31\cdot02+0\cdot17-26\cdot41-13\cdot21)=-0\cdot71 \ tons.$$

The total horizontal load due to earth pressure is resisted by the horizontal reaction of the ground at A and the horizontal resistance of the frame at B. It is the latter force which represents the shear in the frame, and which must be produced by suitable bending moments in the columns. It is, in this instance, equal to one-third of the load (see **5.0**) or $-3\cdot2$ tons, which must be applied.

Of this force $0\cdot71$ tons has been supplied by the moments of the first distribution. There thus remains $-2\cdot49$ tons to be induced by the correcting moments.

Correcting Moments. Since the columns are equal in length and moment of inertia, the arbitrary moments are applied equally to AB and CD and may be taken as 10 ft.-tons top and bottom **(c)**. The distribution of these moments gives an unbalanced shear of

$$\tfrac{2}{12}(-8\cdot01-6\cdot01)=-2\cdot34 \ tons.$$

The unbalanced shear required as a correcting force is $2\cdot49$ tons in the same direction. The moments of **(c)** must be multiplied by $2\cdot49/2\cdot34=1\cdot064$. The following results are obtained:

Moment at	First Moment Distribution (b)	Correcting Moments (c) × by 1·064	Final Moments (d)
AB	+ 0·17	−8·53	− 8·36
BA	+31·02	−6·40	+24·62
CD	−26·41	−6·40	−32·81
DC	−13·21	−8·53	−21·74

It is now possible, knowing the dimensions of the proposed section at any part of the frame, to check the stresses in the steel and concrete—a procedure not within the scope of this book.

6.11

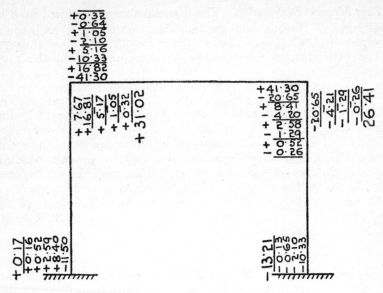

First Moment Distribution

(b)

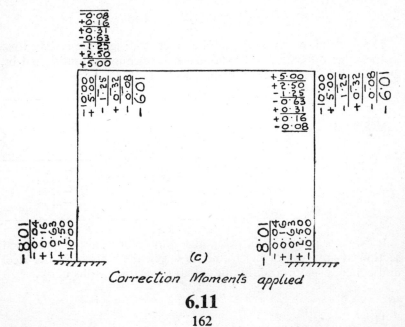

(c)

Correction Moments applied

6.11

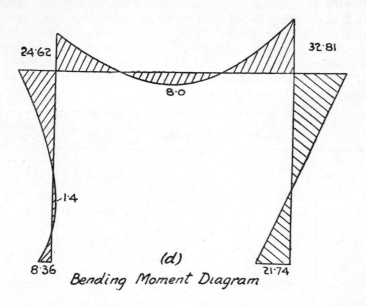

24.62   32.81

8.0

1.4

8.36   21.74

(d)
Bending Moment Diagram

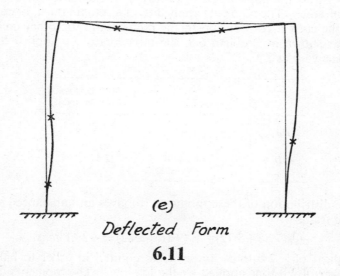

(e)
Deflected Form
6.11

6.12 It is usual to consider the wind forces on a framed building to act as a series of horizontal loads concentrated at the floor levels. Thus, when considering wind forces, none of the beams or columns is loaded transversely, and there is no necessity to determine fixed-end moments or to carry out the first Moment Distribution which was the preliminary step to the solution of problems **6.7** to **6.11**.

Joint	Member	Relative Stiffness	Sum	Distribution Factors
B	BA	$\dfrac{I}{10}=0\cdot10I$	$0\cdot19I$	0·527
	BC	$\dfrac{2\cdot7I}{30}=0\cdot09I$		0·473
C	CB	$\dfrac{2\cdot7I}{30}=0\cdot09I$	$0\cdot24I$	0·375
	CD	$\dfrac{1\cdot8I}{12}=0\cdot15I$		0·625

Horizontal Shear. The solution of this type of problem commences with the application of arbitrary moments to the columns of the frame. The moments applied to one column must, of course, bear the correct relationship to the moments in the other column. As was shown in Problem **6.8**, the moments to be applied to the columns vary as I/l^2.

	I/l^2	Suitable Arbitrary Moment
BA	$\dfrac{I}{100}=0\cdot01I$	10 ft.-tons
CD	$\dfrac{1\cdot8I}{144}=0\cdot0125I$	12·5 ,,

The distribution of these moments (b) gives an unbalanced shear on the frame of

$$-\tfrac{1}{10}(7\cdot85+5\cdot72)-\tfrac{1}{12}(9\cdot16+5\cdot82)=-2\cdot61 \ tons.$$

A shear of $-2\cdot0$ tons, however, is required in order to balance the shear of 2·0 tons applied by the loading. The moments of (b) must therefore be multiplied by 2·0/2·61 to give the values of the final moments. Draw the deflected form of the frame.

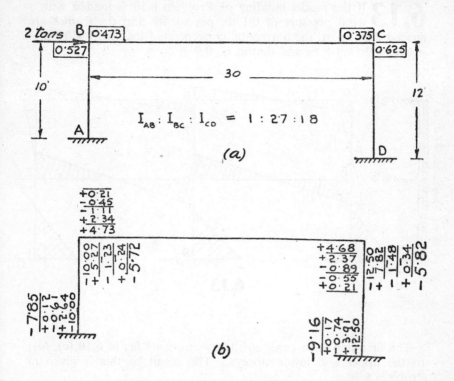

$$I_{AB} : I_{BC} : I_{CD} = 1 : 2{\cdot}7 : 1{\cdot}8$$

(a)

(b)

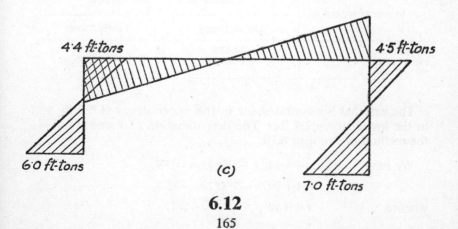

4·4 ft-tons

4·5 ft-tons

6·0 ft-tons

(c)

7·0 ft-tons

6.12

165

6.13 If the model building of Problem **6.10** is loaded with a wind pressure of 0·1 lb. per sq. in. and the frames are spaced 3 in. apart, the horizontal concentrated load at B (**6.10 (a)**) is $6 \times 3 \times 0.1 = 1.8$ *lb.* and that at G, 0·9 *lb.*

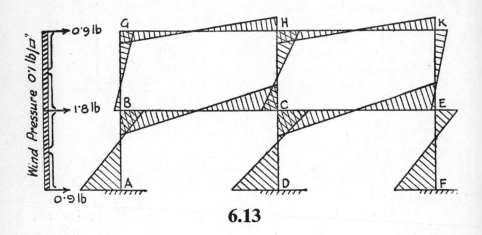

6.13

The first step is to apply arbitrary moments (as in **6.10 (c), (d)**) to the upper and lower storeys. The result of this is given in Problem **6.10** as

Moments applied to	Unbalanced Horizontal Shear in	
	Upper Storey	Lower Storey
Upper storey	−4·642 lb.	+1·897 lb.
Lower storey	+1·963 lb.	−7·352 lb.

The external horizontal shear in the upper storey is 0·9 lb. and in the lower storey 2·7 lb. The determination of x and y follows the method of Problem **6.10**.

We have
$$-4.642x + 1.963y = -0.9$$
$$+1.897x - 7.352y = -2.7,$$
whence
$$x = 0.40 ; \qquad y = 0.47.$$

166

Multiplying the moments obtained in **6.10 (c)** and **6.10 (d)** by x and y respectively, the bending moment diagram shown in Fig. **6.13** is obtained. The table detailing the calculation is given below.

Moment at	Moments from 6.10 (c) × by 0·40	Moments from 6.10 (d) × by 0·47	Final Moments in.-lb.
AB	+0·53	−3·77	−3·24
DC	+0·46	−4·06	−3·60
FE	+0·53	−3·77	−3·24
BA	+1·06	−2·85	−1·79
BG	−1·86	+1·60	−0·26
BC	+0·80	+1·25	+2·05
CH	−2·40	+1·20	−1·20
CE	+0·74	+1·11	+1·85
CD	+0·92	−3·42	−2·50
CB	+0·74	+1·11	+1·85
EK	−1·86	+1·60	−0·26
EF	+1·06	−2·85	−1·79
EC	+0·80	+1·25	+2·05
GB	−1·36	+0·34	−1·02
GH	+1·36	−0·34	+1·02
HG	+1·15	−0·23	+0·92
HK	+1·15	−0·23	+0·92
HC	−2·30	+0·46	−1·84
KH	+1·36	−0·34	+1·02
KE	−1·36	+0·34	−1·02

Draw the deflected form of the frame and compare it with **6.10 (f)**.

Exercises

Solve, by *Moment Distribution*, problems 3.11, 3.12.

6.14 This problem is similar to Problem **5.10** in that a simple consideration of horizontal forces is not sufficient to determine the necessary equation of equilibrium. The inclination of the members AB and CD makes it essential that account should be taken of the vertical reaction V.

Equation of Equilibrium. As in Problem **5.10**, moments must be taken about a sufficient number of points to enable the unknown forces H and V to be eliminated.

Moments about B, $M_{AB} + 2V - 7H = +M_{BC} = -M_{BA}$.

Moments about C, $M_{AB} + 8V - 7H - 30 = +M_{CD}$.

Moments about D, $M_{AB} + 12V + 3H - 70 = -M_{DC}$.

Eliminating H and V between these equations, we have

$$+10M_{AB} + 18M_{BA} + 15M_{CD} + 7M_{DC} = 40$$

which is the required condition of equilibrium with which the final values of the moments must comply.

Distribution Factors.

Joint	Member	Relative Stiffness	Sum	Distribution Factors
B	BA	$\dfrac{2I}{7 \cdot 29} = 0 \cdot 274I$	1·274I	0·215
	BC	$\dfrac{6I}{6} = 1 \cdot 00I$		0·785
C	CB	$\dfrac{6I}{6} = 1 \cdot 00I$	1·278I	0·782
	CD	$\dfrac{3I}{10 \cdot 79} = 0 \cdot 278I$		0·218

First Moment Distribution. This is carried out in the normal way, the result being given in Fig. (c).

Conditions for Sidesway Correction. The frame, loaded as shown in (a), sways laterally, and because of the presence of the inclined members, there is a rotation of BC and not merely a translatory movement along its own line as has been the case in the problems studied previously. The correcting moments for sidesway must be consistent with the rotations shown in (b).

168

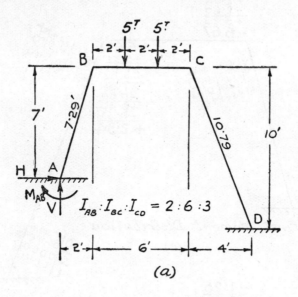

$$I_{AB} : I_{BC} : I_{CD} = 2 : 6 : 3$$

(a)

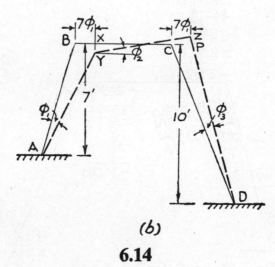

(b)

6.14

169

6.14

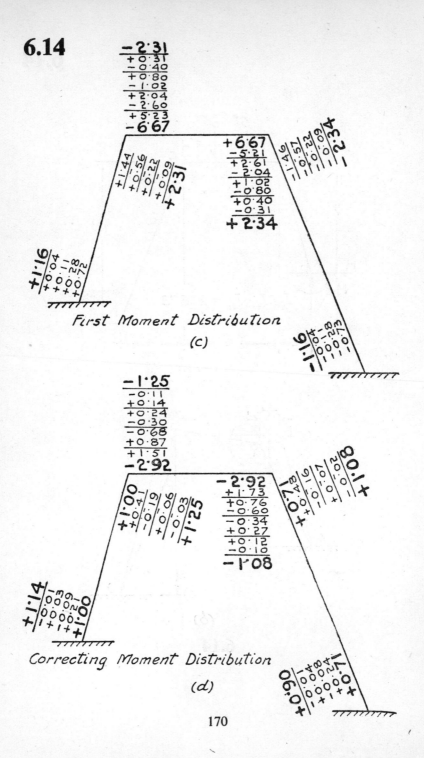

First Moment Distribution

(c)

Correcting Moment Distribution

(d)

170

The horizontal movement of B must equal the horizontal movement of C, i.e. $BX = CP = 7\phi_1$.

Also, $10\phi_3 = CP = 7\phi_1$. Therefore, $\phi_3 = 0.7\phi_1$.

The total relative vertical movement of B and C is

$$= XY + PZ = \tfrac{2}{7} \times 7\phi_1 + \tfrac{4}{10} \times 7\phi_1 = 4.8\phi_1.$$

But $6\phi_2 = XY + PZ = 4.8\phi_1$. Therefore, $\phi_2 = 0.8\phi_1$.

In Problem **6.10** it was shown that for a small displacement of the type shown in **(b)**,

$$M = 6EK\phi.$$

Therefore, $M_{AB} : M_{BC} : M_{CD} = 0.274\phi_1 : \phi_2 : 0.278\phi_3$
and, substituting the values of ϕ_2 and ϕ_3 in terms of ϕ_1, this becomes, on simplification,

$$M_{AB} : M_{BC} : M_{CD} = \phi_1 : 2.92\phi_1 : 0.71\phi_1.$$

Using moments at both ends of AB, BC and CD in these proportions, the second moment distribution is carried out as is shown in **(d)**.

Final Moments. Substituting M_{AB}, M_{BA}, M_{CD} and M_{DC} from the first Moment Distribution **(c)** in the Equation of Equilibrium, we have

$$+10M_{AB} + 18M_{BA} + 15M_{CD} + 7M_{DC} = 40$$
$$\text{but,} \quad +11.6 + 41.6 - 35.1 - 8.1 = 10.$$

The right-hand side of the equation should be 40, which means that 30 must be supplied by the correcting moments.

Substituting M_{AB}, M_{BA}, M_{CD} and M_{DC} from the second Moment Distribution in the Equation of Equilibrium, we have

$$+11.4 + 22.5 + 16.2 + 6.3 = 56.4.$$

Actually, only 30 is required. The moments obtained in the second distribution must therefore be multiplied by $30/56.4 = 0.532$, and added to the moments from **(c)**. The following table gives the final results:

Moment at	First Moment Distribution (c)	Correcting Moment Distribution (d)	Correcting Moments multiplied by 0.532	Final Moments: ft.-tons
AB	+1.16	+1.14	+0.61	+1.77
BA	+2.31	+1.25	+0.66	+2.97
CD	−2.34	+1.08	+0.57	−1.77
DC	−1.16	+0.90	+0.48	−0.68

6.14

Draw the bending moment diagram and the deflected form of the frame. Determine also the maximum bending moment on the frame, which occurs at the centre of the beam *BC*.

The problems in this chapter have been deliberately kept short and simple in order to demonstrate clearly the method employed. Results in closer agreement with the " exact " solutions will be obtained by working within closer limits and by carrying out further balancing and distribution operations. The early problems of the chapter, for example, should be solved by the student to two places of decimals.

Revision Exercises

Solve these problems by *Moment Distribution*:

A cantilever is 12 ft. long and is propped at the free end to the same level as the fixed end. A concentrated load of 5 tons is carried by the cantilever at 2 ft. from the propped end. What is the load on the prop? (3·76 *tons*.)

A continuous beam has two spans *AB* and *BC* and is freely supported at *A* and *C*. *AB*=45 ft. : *BC*=36 ft. *AB* carries a uniformly distributed load of 2 tons per ft. over its whole length. *BC* carries a concentrated load of 20 tons at 15 ft. from *C*. ($M_B = -605$ *ft.-tons*.)

A continuous beam *ABCD* is fixed horizontally at *A*, simply supported at B and *C* and cantilevered from *C* to *D*. *AB*=*BC*=10 ft. : *CD*=5 ft. There is a U.D.L. of 2 tons per ft. over *BC*, and a concentrated load of 1 ton at *D*. ($M_B = -12·8$ *ft.-tons*.)

CHAPTER 7

THE COLUMN ANALOGY *

The analogy is that which exists between the equations for slopes and deflections in a bent beam, and for loads and moments in a short column eccentrically loaded. It is assumed that these simple expressions are familiar, and the analogy on which the method is based can be appreciated by a simple comparison.

Equations for a Bent Beam

Angle between tangents,

$$\phi = \int m_i \frac{ds}{EI}$$

Vertical deflection of one end from the tangent at the other end

$$\Delta_y = \int m_i x \frac{ds}{EI}$$

Horizontal deflection of one end from the tangent at the other end

$$\Delta_x = \int m_i y \frac{ds}{EI}$$

m is the indeterminate moment at any point caused by restraint. The "static" moments caused by the external loading on the beam when the latter is in a statically determinate condition (resulting from the release of restraints) must be accounted for separately. The rotation of a fixed end is zero, and thus ϕ must be counteracted by an opposite rotation caused by the loading on the beam when it is in the statically determinate condition. This rotation $= \int m_s \frac{ds}{EI}$ where m is the "static" moment at any section.

Equations for a Short Column eccentrically loaded

Total load on the column, $P = \int p.dA.$

Moment about the Y axis, $M_y = \int px.dA$

Moment about the X axis, $M_z = \int py.dA.$

p is the fibre stress at any point of the column section.

* Reprinted from *The Structural Engineer*, August, 1939.

The Analogy

A comparison of the two sets of equations shows that $\int \dfrac{ds}{EI}$ in a bent beam corresponds to $\int dA$ of a short column, and $\int m_s \dfrac{ds}{EI} = \phi$ in a bent beam corresponds to P of a short column ; m_i in a bent beam corresponds to p of a short column.

Thus the analogous column corresponding to a beam must fulfil the following conditions:

(1) The length of the cross-section of the column should be equal to the length of the beam, and if the beam is not straight the shape of the cross-section of the column should follow the shape of the side of the beam.

(2) The width of the column cross-section at any point should be equal to the value of $1/EI$ of the corresponding section of the beam.

From (1) and (2) the cross-section of the column has an area equal to $\int \dfrac{ds}{EI}$ of the beam.

Figs. *7.1 et seq.* show beams with their analogous columns.

(3) The column should be loaded with an intensity equal at any section to the "static" bending moment at a corresponding point on the beam (m_s).

This bending moment is found by releasing statically indeterminate moments and forces until the beam is in a statically determinate condition under the given loading. For any beam and loading there may be a number of such statically determinate conditions, any of which may be used. This conception of a bending moment diagram as a "load" is familiar in the "area-moment" method for beams. The term m_s used above is the ordinate of this bending moment diagram at any point.

If the above conditions are fulfilled, the fibre stress at any point on the analogous column section is numerically equal to the indeterminate moment (produced by restraint) at the corresponding section of the beam.

Convention of Signs

Positive bending moment produces tension on the underside of a beam or the inside of a frame.

Positive co-ordinates are measured upwards and to the right of the origin.

Column Analogy, like Moment Distribution, shows to best advantage in its simplest form. Although not of such wide application as Moment Distribution, it has certain uses beyond the scope of the present discussion, but to the designer Column Analogy has its most useful application in the rapid analysis of fixed base, symmetrical and unsymmetrical portal frames and arches.

BEAMS

7.1 (a) shows a built-in beam of uniform moment of inertia It is required to determine the fixing moments.

(b) shows the analogous column whose length equals that of the beam and whose width is constant and equal to $1/EI$ of the beam.

To determine the applied loading on the analogous column it is necessary to reduce the beam to a statically determinate condition. Three of such possible conditions, together with their bending moment diagrams, are shown in (d), (e), (f).

(c) illustrates the analogous column loaded eccentrically with one of these diagrams, the intensity of the "loading" varying as the bending moment on the simply supported beam.

The next procedure is the finding of the fibre stresses in the column at the ends A and B.

Area of cross-section $(A) = 10/EI$.

Total load $(P = m_sA)$ $= 9 \cdot 6/2 \times 10/EI = 48/EI$.

Moment of inertia $= \dfrac{\dfrac{1}{EI} \times 10^3}{12} = \dfrac{83 \cdot 3}{EI}$

Centroid of column section is 5 ft. from each end.

Resultant of total load is 6 ft. from B.

Eccentricity of load, 1 ft.

Stress at the ends A and B

$$= \frac{P}{A} \pm \frac{M_y x}{I} = \frac{48}{EI} \times \frac{EI}{10} \pm \frac{\left(\frac{48}{EI} \times 1\right)5}{\frac{83 \cdot 3}{EI}} = 7 \cdot 68 \text{ at } A \text{ and } 1 \cdot 92 \text{ at } B.$$

These are the "indeterminate" moments in ft.-tons at A and B respectively. The final moment at any section is equal to the "static" moment less the "indeterminate" moment.

$$M_A = 0 - 7 \cdot 68 = -7 \cdot 68 \text{ ft.-tons.}$$
$$M_B = 0 - 1 \cdot 92 = -1 \cdot 92 \quad \text{,,}$$

7.1

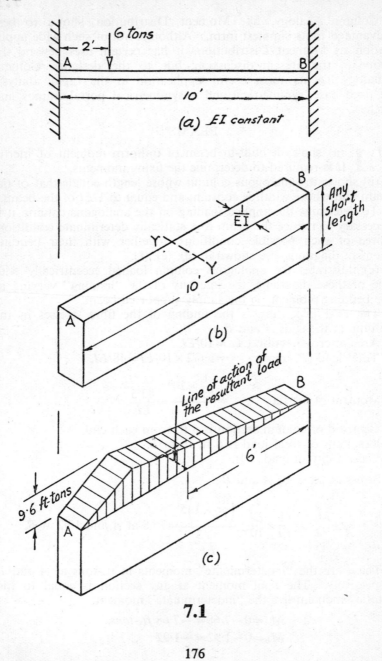

6 tons

A ⟶ 2' ⟶ B

⟵ 10' ⟶

(a) EI constant

B

Any short length

1/EI

Y
Y

10'

A

(b)

Line of action of the resultant load

B

9.6 ft-tons

6'

A

(c)

7.1

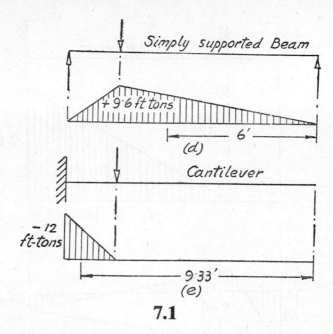

(d)

(e)

The evaluation of the indeterminate fixing moments can be carried out with either of the other two bending moment diagrams (Fig. **7.1**). Using the second diagram, the mean bending moment (m_s) is -6 ft.-tons and the area over which the load acts is $2/EI$.

Total load $(P) = -12/EI$.

Eccentricity of load, $9.33 - 5.0 = 4.33$ ft.

Stress at A and B

$$= \frac{P}{A} \pm \frac{M_y x}{I} = -\frac{12}{EI} \times \frac{EI}{10} \pm \frac{\left(\frac{12}{EI} \times 4.33\right)5}{\frac{83.3}{EI}} = -4.32 \text{ or } +1.92.$$

Final moment, $m_s - m_i$.

At ends $M_A = -12 - (-4.32) = -7.68$ ft.-tons.

$$M_B = 0 - 1.92 = -1.92 \quad ,,$$

(g) and (h) show the construction of the bending moment diagrams from the data given by these two calculations.

7.1

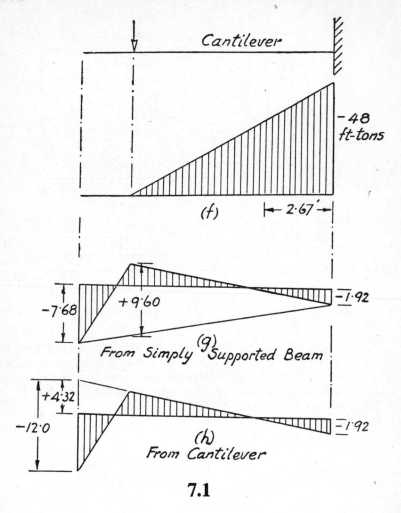

Cantilever

−48 ft-tons

|← 2·67′ →|

(f)

−7·68

+9·60

−1·92

(g)
From Simply Supported Beam

+4·32

−12·0

−1·92

(h)
From Cantilever

7.1

Exercises

Solve, by *Column Analogy*, problems 3.6, 3.3, 3.4, 3.5, 3.7, 5.8, 6.5, 6.7, 6.8, 6.11.

7.2 A built-in beam is composed of two lengths having different moments of inertia. As shown by the free bending moment diagram, the average "static" moment on each section is 16 ft.-tons.

Centroid of Column Section.

Left Section—

    Cross-sectional area (A)   = 4/2EI = 2/EI.

    Moment of area about B = 20/EI.

Right Section—

    Cross-sectional area (A)   = 8/EI.

    Moment of area about B = 32/EI.

    Centroid of column section is 52/EI × 10/EI = 5·2 ft. from B.

Moment of Inertia of Column Section about an axis YY through the centroid

$$= \left\{ \left(\frac{\frac{1}{2} \times 4^3}{12} + 2 \times 4 \cdot 8^2 \right) + \left(\frac{1 \times 8^3}{12} + 8 \times 1 \cdot 2^3 \right) \right\} \frac{1}{EI} = \frac{102 \cdot 9}{EI}$$

A simpler method of determining the moment of inertia is to calculate its value about some arbitrary axis, such as one through B, and then to make a correction to the axis through the centroid. In this instance the correction to the centroid is

$$\text{Total area} \times 5 \cdot 2^2 = \frac{10}{EI} \times 5 \cdot 2^2 = \frac{270 \cdot 4}{EI}$$

This correction is always subtracted.

Bending Moment.

Left Section—

Eccentricity of load: 4·13 *ft.*

$$P \times 4 \cdot 13 = m_s A \times 4 \cdot 13 = (16 \times 2 \times 4 \cdot 13) \frac{1}{EI}$$

Right Section—

$$P \times 0 \cdot 13 = m_s A \times 0 \cdot 13 = (16 \times 8 \times 0 \cdot 13) \frac{1}{EI}$$

Total moment = 148·8/EI.

Fibre Stress.

$$\text{at} \quad A = \frac{P}{A} + \frac{M_y x}{I} = \frac{106}{10} + \frac{148 \cdot 8 \times 6 \cdot 8}{102 \cdot 9} = 25 \cdot 9$$

$$\text{at} \quad B = \frac{160}{10} - \frac{148 \cdot 8 \times 5 \cdot 2}{102 \cdot 9} = 8 \cdot 5.$$

Since there is no static moment at A or B the final moments are −25·9 and −8·5 ft.-tons respectively.

7.2

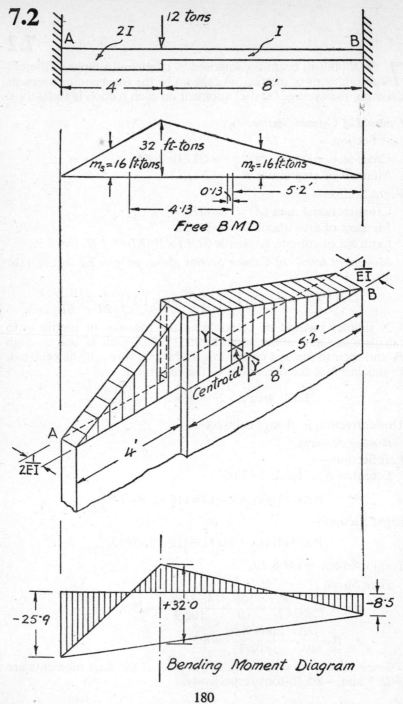

12 tons

A   2I   I   B

4'   8'

32 ft-tons

$m_s = 16$ ft-tons $m_s = 16$ ft-tons

0.13   5.2'

4.13

Free BMD

$\dfrac{1}{EI}$ B

Y

Centroid   5.2'

8'

A

$\dfrac{1}{2EI}$ 4'

−25.9   +32.0   −8.5

Bending Moment Diagram

There is, of course, no advantage in using the method of Column Analogy for such simple problems as those investigated above, and when extended to more difficult beams or frames the above operations become complicated. It is better, therefore, to carry out the work in tabular form, and in further examples this presentation will be used. To assist the reader, the calculations for Problem 7.2 are set out in a table.

The quantity EI for the beam is constant and has been given a relative value of unity.

From columns 4 and 5 the centroid of the column section is found to be 5·2 ft. from B. From columns 6 and 7 the total I for the column is 102·9 units, and the remainder of the calculation is the same as for the first derivation.

1	2	3	4	5	6	7	8	9	10
Section	Length (d)	Analogous Column Width (b)	Analogous Column Area (A)	Moment of Area about B (Ax)	Second Moment of Area about B (Ax^2)	Moment of Inertia ($I_0 = bd^3/12$)	Mean B.M. (m_s)	Load ($P = m_s A$)	Moment of Load about Centroid (M_y)
Left .	4	$1/2EI$	2	20	200	2·6	16	32	$32 \times 4·13$
Right .	8	$1/EI$	8	32	128	42·7	16	128	$128 \times 0·13$
			10	52	328	45·3			148·8

Correction to Centroid 270·4

Nett Ax^2 57·6

7.3 The portal frame, for the purpose of this problem, will be considered to be in the two-hinged condition. A hinge represents the possibility of an indefinitely large rotation. This results in the following:

(1) The area of the analogous column cross-section corresponding to the hinges of the frame is infinite.

(2) The centroid of the column section lies midway between the hinges.

(3) The moment of inertia of the column cross-section about YY is infinite.

(4) The neutral axis of the column cross-section passes through both hinges.

In these circumstances the fibre stress which is normally

$$= P/A \pm M_y x/I_y \pm M_x y/I_x$$

becomes $\qquad = \pm M_x y/I_x.$

(b) shows the loading on the analogous column. EI is again constant and has been given a relative value of unity.

Stress at B and $C = 144 \times 6/432 \cdot 0 = 2$.

The indeterminate moments at B and C are caused by the horizontal force H at the hinges and are each -2 ft.-tons. The final bending moment diagram is obtained by superimposing the static and indeterminate diagrams in the usual way.

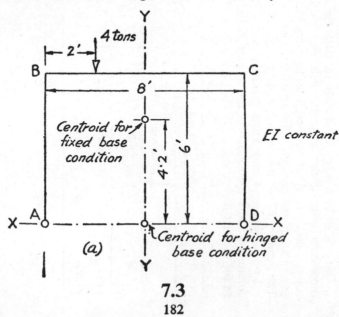

7.3

Hinged base condition

1	2	3	4	5	6	7	8	9
Member	Length (d)	Vertical Height from Origin (y)	Analogous Column Area (A)	Second Moment of Area about N.A. (Ay^2)	Moment of Inertia ($I=bd^3/12$)	Mean B.M. (m_s)	Load ($P=m_sA$)	Moment of Load about Centroid (M_x)
Right leg . .	6	3	6	54	18·0	—	—	—
Beam . .	8	6	8	288	—	3	24	144
Left leg . .	6	3	6	54	18·0	—	—	—
				396	36·0			144

36·0 : Total moment of inertia = 432·0

Fixed base condition

1	2	3	4	5	6	7	8	9	10	11
Member	Length (d)	Vertical Distance from Origin (y)	Horizontal Distance from Origin (x)	Analogous Column Area (A)	Moment of Area about XX axis (Ay)	Second Moment of Area about YY axis (Ax^2)	Second Moment of Area about XX axis (Ay^2)	I_{ox}*	I_{oy}*	Load ($P=m_sA$)
Right leg .	6	3	+4	6	18	96	54	18·0	—	—
Beam . .	8	6	0	8	48	—	288	—	42·7	24
Left leg . .	6	3	−4	6	18	96	54	18·0	18·0	—
				20	84	192	396	36·0	42·7	

* I_{ox} and I_{oy} signify the moments of inertia of the various members about axes through their respective centroids.

7.3

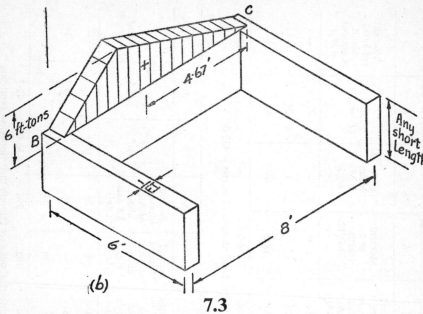

(b)

7.3

Fixed base condition

The frame will now be considered to be fixed at the bases of the legs. The centroid of the column cross-section is now no longer on the line of the hinges, since the infinite area representing those hinges vanishes. The bending stresses about both the XX and the YY axes must be determined. These axes are arbitrarily chosen and a correction made to the centroid at a later stage. The origin is at the centre of AD as before.

Position of centroid $=84/20=4\cdot2\,ft.$ above XX.

Correction to centroid $=20\times4\cdot2^2=352\cdot8$.

Total $I_{xx}=396+36\cdot0-352\cdot8=79\cdot2$

Total $I_{yy}=192+42\cdot7=234\cdot7$.

Eccentricity of load about the centroid is $1\cdot8$ ft. in the Y direction and $-0\cdot67$ ft. in the X direction.

$$M_A=\frac{24}{20}-\frac{(24\times1\cdot8)\times4\cdot2}{79\cdot2}+\frac{(24\times0\cdot67)\times4}{234\cdot7}=-0\cdot82\,ft.\text{-}tons.$$

Similarly $M_B=2\cdot44\,ft.\text{-}tons$ $M_C=1\cdot90\,ft.\text{-}tons$

and $M_D=-1\cdot34\,ft.\text{-}tons.$

Since there are no static moments at these points, the above figures, with opposite signs, give the values of the final bending moments at A, B, C and D.

7.4 This frame is considered to be in the "fixed base" condition.

Area of free bending moment diagram

$$=\left(\frac{150}{4}-25\right)2\cdot5+\int_0^5\left(\frac{30}{4}x-x^2\right)dx=83\cdot4$$

$m_s=8\cdot34.$

Moment of diagram about C

$$=31\cdot25\times6\cdot67+\int_0^5\left(\frac{30}{4}x^2-x^3\right)dx=364\cdot7.$$

Centroid of the area is $364\cdot7/83\cdot4=4\cdot4$ ft. from C.
Correction to centroid $=29\times7\cdot03^2=1435.$

$$M_D=\frac{41\cdot7}{29}-\frac{(41\cdot7\times4\cdot97)\times7\cdot03}{437}+\frac{(41\cdot7\times0\cdot62)\times5}{641\cdot7}.$$

Indeterminate $M_D=-1\cdot69$ ft.-tons.

    Final $M_D=+1\cdot69$ ft.-tons.

Determination of Neutral Axis. The line of the neutral axis of the analogous column section determines the points of contraflexure in the frame, and thus enables the bending moment diagram for the frame to be drawn.

$$Stress=P/A+M_xy/I_x+M_yx/I_y.$$

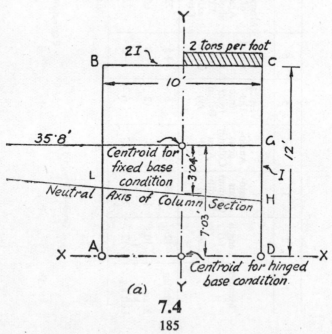

(a)

1	2	3	4	5	6	7	8	9	10	11	12	13
Member	Length (d)	Analogous Column Width (b)	Analogous Column Area (A)	Vertical Distance from Origin (y)	Horizontal Distance from Origin (x)	Moment of Area about XX axis (Ay)	Second Moment of Area about YY axis (Ax)	Second Moment of Area about XX axis (Ay²)	I_{ox}	I_{oy}	Mean B.M. (m_v)	Load ($P = m_s A$)
Right leg	12	1/EI	12	6	+5	72	300	432	144		—	—
Beam .	10	1/2EI	5	12	0	60	—	720	—	41·7	8·34	41·7
Left leg .	12	1/EI	12	6	−5	72	300	432	144		—	—
			29			204	600	1584 288 ‾‾‾‾ 1872 1435 ‾‾‾‾ 437	288	41·7 600·0 ‾‾‾‾ 641·7		41·7

Correction to centroid

$\bar{y} = 204/29 = 7\text{·}03$ ft.

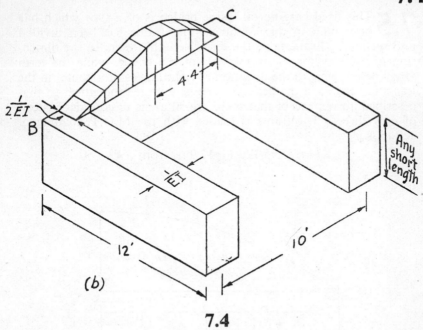

(b)

7.4

Equating stress to zero and also x and y to zero in turn,

$$y_1 = -\frac{P/A}{M_x/I_x} \text{ and } x_1 = -\frac{P/A}{M_y/I_y}.$$

These are the intercepts of the neutral axis on the X and Y axes through the centroid. In this example:

$$y_1 = -\frac{41\cdot7}{29} \times \frac{437}{41\cdot7 \times 4\cdot97} = -3\cdot04 \text{ ft.}$$

$$x_1 = -\frac{41\cdot7}{29} \times \frac{643\cdot7}{41\cdot7 \times 0\cdot62} = -35\cdot8 \text{ ft.\}$$

The neutral axis is shown in Fig. (a). By scaling or by calculation, HD can be determined.

Then horizontal force (H) at the hinge $\times HD + M_D = 0$.

$$\therefore H = \frac{1\cdot69}{HD} = \frac{1\cdot69}{3\cdot55} = 0\cdot47 \text{ tons.}$$

L and H are points of contraflexure. 7.4 (a)

7.5 This fixed base portal frame bridge is of a type which has been built in this country. The frame has been divided into sections. In the table the term "length" refers to the dimension of each section along the axis of the frame, while the term "thickness" refers to the direction at right angles (horizontal in the legs and vertical in the beam). The procedure is identical to that described above, except that there are 30 areas to consider instead of only three. The frame is loaded with the M.O.T. loading of 220 lb./sq. ft.

$$y = 1553 \cdot 5 / 120 \cdot 1 = 12 \cdot 9 \ ft. \ \text{from } XX.$$

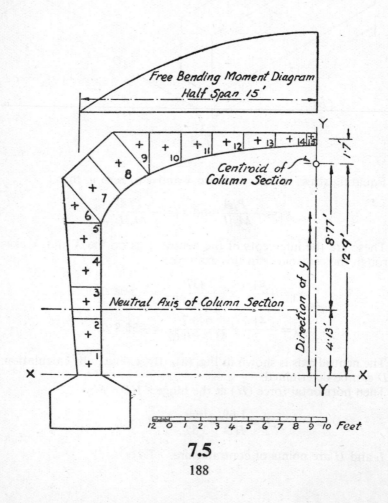

7.5

1	2	3	4	5	6	7	8	9	10
Section	Length (d) ft.	Thickness (t) ft.	Moment of Inertia $(t^3/12)$	Analogous Column Area $(A=12d/t^3)$	Vertical Distance from Origin (y)	Moment of Area about XX axis (Ay)	Second Moment of Area about XX axis (Ay^2)	Mean B.M. (m_s)	Load $(P=m_s A)$
1	1·5	1·7	0·39	3·8	0·7	2·7	1·9	—	—
2	2·0	1·8	0·45	4·5	2·4	10·8	25·9	—	—
3	2·0	1·9	0·55	3·6	4·4	15·8	69·7	—	—
4	2·0	2·0	0·67	3·0	6·4	19·2	122·8	—	—
5	2·0	2·2	0·90	2·2	8·4	18·5	155·2	—	—
6	1·3	2·3	0·95	1·4	10·3	14·4	148·5	—	—
7	2·1	3·3	3·00	0·7	11·7	8·2	95·8	—	—
8	2·1	3·2	2·73	0·8	13·1	10·5	137·3	0·83	0·67
9	1·1	2·6	1·43	0·8	14·4	11·5	165·9	1·18	0·94
10	2·0	2·2	0·90	2·2	14·0	30·8	431·0	1·64	3·61
11	2·0	1·6	0·36	5·8	14·3	82·9	1186·0	2·04	11·83
12	2·0	1·3	0·18	11·2	14·4	161·3	2323·0	2·34	26·20
13	2·0	1·0	0·08	24·1	14·5	349·4	5067·0	2·59	62·41
14	2·0	0·8	0·05	41·7	14·6	608·8	8890·0	2·74	114·20
15	0·5	0·8	0·04	14·3	14·6	208·7	3049·0	2·76	39·47
For half frame only				120·1		1553·5	21,869 / 20,080 / 1789		259·33

189

7.5

The moment of inertia of the column cross-section (column 8) is obtained about the XX axis. A correction must be made to a parallel axis through the centroid after the position of the latter has been determined. This correction is $Ay^2 = 20,080$ and is subtracted. In considering the loading (P) on the column and the moment of inertia of the column cross-section it must be remembered that the frame is symmetrical about YY.

Fixing moment at the base

$$= -\left(\frac{P}{A} + \frac{M_x y}{I_x}\right) = -\left(\frac{2 \times 259 \cdot 3}{240} + \frac{(2 \times 259 \cdot 3 \times 1 \cdot 7)(-12 \cdot 9)}{3578}\right)$$
$$= -2 \cdot 16 + 3 \cdot 18 = +1 \cdot 02 \ ft.\text{-}tons.$$

The position of the neutral axis can be found by equating the bending moment to zero.

$$\therefore y \frac{M_x}{I_x} = -\frac{P}{A}.$$

$$y = -\frac{P/240}{1 \cdot 7 P/3578} = -\frac{3578}{1 \cdot 7 \times 240} = -8 \cdot 77 \ ft.$$

Since the frame is symmetrical about YY, the neutral axis is parallel to XX and 4 ft. $1\frac{1}{2}$ in. above the axis XX.

UNSYMMETRICAL FRAMES

The three portal frame problems discussed above have dealt with frames which are symmetrical about the YY axis. Such frames are in the majority, but the following illustrates the method of dealing with frames which are unsymmetrical about both axes.

Stress at any Point on a Column Cross-section

Symbols:

b, c and d are constants.

I_x is the moment of inertia of the column cross-section about the X axis through the centroid.

I_y is the moment of inertia of the column cross-section about the Y axis through the centroid.

I_{xy} is the product moment about the X and Y axes through the centroid.

P is the direct force. p is the unit stress at any point.

M_x is the bending moment about the X axis and $= Py$.

M_y is the bending moment about the Y axis and $= Px$.

Stress at any Point on a Column Cross-section

$$p = b + cx + dy \quad . \quad . \quad . \quad . \quad . \quad . \quad . \quad . \quad . \quad (1)$$

$$P = \int p\,dA = b\int dA + c\int x\,dA + d\int y\,dA.$$

Axes are through the centroid

$$\therefore \ x\,dA = y\,dA = 0.$$

$$\therefore \ b = P/A \quad . \quad . \quad . \quad . \quad . \quad . \quad . \quad . \quad . \quad (2)$$

$$M_x = \int py\,dA = b\int y\,dA + c\int xy\,dA + d\int y^2\,dA = cI_{xy} + dI_x \ . \quad (3)$$

$$M_y = \int px\,dA = b\int x\,dA + c\int x^2\,dA + d\int xy\,dA = cI_y + dI_{xy} \ . \quad (4)$$

Multiply by I_{xy}/I_y and eliminate c from equations (3) and (4).

$$d = \frac{M_x - M_y\dfrac{I_{xy}}{I_y}}{I_x - \dfrac{I_{xy}^2}{I_y}} = \frac{M'_x}{I'_x}$$

Similarly

$$c = \frac{M_y - M_x\dfrac{I_{xy}}{I_x}}{I_y - \dfrac{I_{xy}^2}{I_x}} = \frac{M'_y}{I'_y}$$

From equation (1) $\qquad p = \dfrac{P}{A} + \dfrac{M'_y}{I'_y}x + \dfrac{M'_x}{I'_x}y$

Equation of neutral axis, where $p = 0$ is

$$\frac{P}{A} + \frac{M'_y x}{I'_y} + \frac{M'_x}{I'_x}y = 0.$$

Put $x = 0$ and $y = 0$ and obtain intercepts.

$$x_1 = -\frac{P/A}{M'_y/I'_y} \quad \text{and} \quad y_1 = -\frac{P/A}{M'_x/I'_x}$$

These are the intercepts of the neutral axis on the XX and YY axes through the centroid of the column section. The points in which the neutral axis cuts the column section indicate the positions of points of zero moment (or points of contraflexure) in the frame.

191

7.6 The calculations for this unsymmetrical portal frame are carried out in a similar manner to those in the last two examples, but the centroid must be located in two directions, and the values of I_x and M_x are replaced by the values of I'_x and M'_x.

The relative values of the moments of inertia of the members of the frame are:

Left leg : Beam : Right leg $=1 : 2 : 3$.

The origin of the two axes XX and YY is at D (Fig. **7.6**).

Position of Centroid.

$$\bar{x}=-\frac{75}{14}=-5\cdot36\,ft. \qquad \bar{y}=\frac{131\cdot5}{14}=9\cdot39\,ft.$$

Correction to Axes through the Centroid.

$$A\bar{x}^2=14\times5\cdot36^2=\ 402\cdot3.$$
$$A\bar{y}^2=14\times9\cdot39^2=1234\cdot0.$$
$$A\overline{xy}=-14\times5\cdot36\times9\cdot39=-704\cdot7.$$

Correction for Dissymmetry.

$$M'_x=M_x-M_y\frac{I_{xy}}{I_y}=217+30\frac{70\cdot3}{264\cdot4}=225\cdot0.$$

$$M'_y=M_y-M_x\frac{I_{xy}}{I_x}=30+217\frac{70\cdot3}{139\cdot7}=139\cdot1.$$

$$I'_x=I_x-\frac{I_{xy}^2}{I_y}=139\cdot7+\frac{70\cdot3^2}{264\cdot4}=121\cdot0.$$

$$I'_y=I_y-\frac{I_{xy}^2}{I_x}=264\cdot8+\frac{70\cdot3^2}{139\cdot7}=229\cdot4.$$

$$\frac{M'_y}{I'_y}=0\cdot606. \qquad\qquad \frac{M'_x}{I'_x}=1\cdot860.$$

$$\frac{P}{A}=\frac{250}{42}=5\cdot95.$$

Intercepts of Neutral Axis on XX and YY Axes through Centroid.

$$x_1=-\frac{5\cdot95}{0\cdot606}=-9\cdot82\,ft.$$

$$y_1=-\frac{5\cdot95}{1\cdot86}=-3\cdot20\,ft.$$

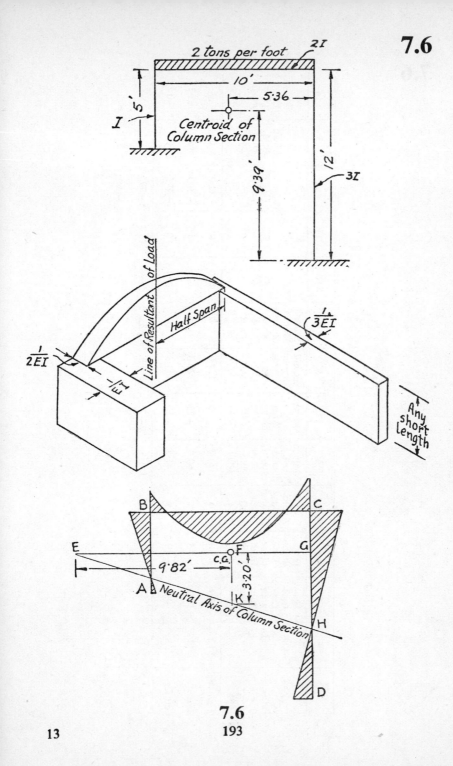

2 tons per foot

2I

10'

5'

5·36

I

Centroid of
Column Section

9·39'

12'

3I

Line of Resultant of Load

Half Span

$\frac{1}{3EI}$

$\frac{1}{2EI}$

$\frac{1}{EI}$

Any short length

B

C

E

F

G

9·82'

c.g.

3·20'

A

K

Neutral Axis of Column Section

H

D

7.6

1	2	3	4	5	6	7	8	9	10	11	12	13	14
Member	Length	Vertical Distance from Origin (y)	Horizontal Distance from Origin (x)	Analogous Column Area (A)	Moment of Area about XX axis (Ay)	Moment of Area about YY axis (Ax)	Second Moment of Area about XX axis (Ay^2)	Second Moment of Area about YY axis (Ax^2)	I_{ox}	I_{oy}	I_{xy} (Axy)	Mean B.M. (m_s)	Load $(P = m_s A)$
Right leg	12	6·0	0	4	24·0	0	144·0	0	48·0	0	0	0	0
Beam	10	12·0	−5	5	60·0	−25	720·0	125	0	41·7	−300	16·7	83·3
Left leg	5	9·5	−10	5	47·5	−50	451·3	500	10·4	0	−475	0	0
				14	131·5	−75	1315·3	625	58·4	41·7	−775		

194

CHAPTER 8

INFLUENCE LINES FOR CONTINUOUS STRUCTURES

For the understanding of this chapter, the student should be familiar with influence lines for statically determinate structures, such as simple beams and frames, and should have clearly in mind the distinction between (for example) a shear force or bending moment diagram, and a shear force or bending moment influence line.

As each problem is studied, reference should be made to the following important statements. (8.0)

(1) An Influence Line is drawn to show the value of a given quantity at one particular section (*XX*) of the structure.

(2) The value of the quantity in question (at section *XX*), when a unit load is at some section *YY*, is represented by the ordinate of the Influence Line Diagram (*zz*) at *YY*.

As a first step towards drawing the Influence Line Diagrams for shear, bending moment, deflection, etc., in a continuous structure, the methods of previous chapters might be used in order to obtain general expressions defining the values of the redundant forces and moments as a unit load traverses the structure. These general expressions are the equations of the Influence Line Diagrams for the redundant quantities. When such Influence Lines have been drawn, the Influence Lines for other quantities can be determined by the methods of statics.

Application of Maxwell's Theorem of Reciprocal Displacements

The methods of Chapters 2 to 7 are satisfactory only when the problem is concerned with a single combination of loads. When the load may occupy many different positions on the structure, solution by means of these methods is clumsy and tedious. The general expressions for the statically indeterminate forces and moments can, in most instances, be obtained more neatly by an application of Maxwell's Theorem of Reciprocal Displacements.

In Fig. 8.0 *XX* is the line of action of a redundant reaction on a continuous structure, and *YY* is the line of action of the unit influence load.

8.0

Suppose, when the redundant reaction is removed, the structure becomes statically determinate. Then the displacement of the beam at XX can be calculated and found equal to δ.

The value of the reaction R, then, must be such that the upward displacement at XX caused by R is equal to the downward displacement at XX caused by a unit load acting at any section YY.

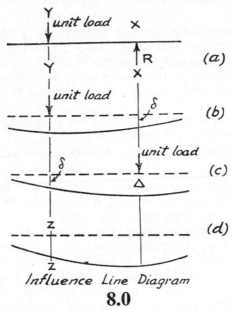

Influence Line Diagram

8.0

Suppose a unit load is placed at XX and the displacement there is found to be Δ, then the value of R for the first position of the unit load is such that $R\Delta = \delta$ or $R = \delta/\Delta$.

But, by Maxwell's Theorem, if a unit load at YY causes a displacement of δ at XX, then a unit load at XX causes the same displacement, δ, at YY. **(c)**

Thus the required values of δ and therefore of δ/Δ, for *all* positions of the unit influence load, can be found from a *single* position of a unit load (at XX) and the influence line for the redundant reaction R is of the same shape as the elastic curve of the structure in **(c)**.

A similar method can be used when the statically indeterminate quantity is a moment. A unit moment is applied at the correct section, and the displacements considered are rotations instead of deflections.

8.1

INFLUENCE LINES FOR STATICALLY INDETER-MINATE BEAMS

8.1 The problem is to determine the influence line for R, the redundant reaction, showing how this quantity varies as the unit load moves across the span.

When the redundant reaction is removed, the unit load at XX causes a certain downward deflection at M. The value of the unknown reaction must be such that it counteracts this deflection by the lifting of M to its original position.

In order to construct the required influence line the student might place the unit load at various sections, and calculate R for each position by the method of Problem **2.1.**

This procedure, however, is clumsy and laborious, and a shorter treatment can be obtained by the use of Maxwell's Theorem of Reciprocal Displacements, as explained above. A unit load is placed at M, and the deflected form of the beam so loaded is of the same shape as the influence line for R. The ordinate of the deflected shape is δ with a maximum deflection of $\varDelta$ at M. The ordinate of the influence line for R is $\delta/\varDelta$ with a maximum value of unity at M.

The expression for the deflection δ is given by the moment of the shaded area of the M/EI diagram (see Chapter 2) about XX.

$$\delta = -\left\{\frac{a(l-a)}{2}\ \frac{(l-a)}{3}+\frac{l(l-a)}{2}\ \frac{2(l-a)}{3}\right\}\frac{1}{EI}=\frac{1}{6EI}(-a^3-2l^3+3al^2).$$

The value of $\varDelta$ is the value of δ when $a=0$.

$$\varDelta = -\frac{l^3}{3EI}$$

$$\therefore R=\delta/\varDelta=\tfrac{1}{2}(2+K^3-3K), \text{ where } K=a/l.$$

A systematic and orderly method of obtaining the ordinates of influence lines of the types mentioned in this chapter is by tabulating the values of the various terms as is shown below.

K	$-3K$	K^3	R tons
0	0	0	+1·00
0·2	−0·60	+0·01	+0·70
0·4	−1·20	+0·06	+0·43
0·6	−1·80	+0·22	+0·21
0·8	−2·40	+0·51	+0·06
1·0	−3·00	+1·00	0

8.1

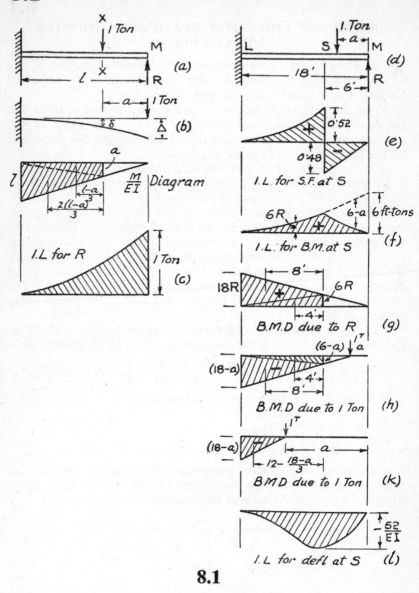

8.1

In (c) the ordinate under the unit load when the latter occupies any position on the beam represents the value of the reaction R for that position of the loading.

From this influence line others can be constructed. Suppose $l=18$ ft. (d) The influence line found above can now be used to determine the influence lines for shear, bending moment and deflection at any point on the beam.

Influence Line for Shearing Force at 6 ft. from M. Lay a piece of paper over (d), covering the length of the beam from L to S. Then it can be seen that, so long as the unit load is to the left of S, the shearing force at S is merely the value of R. When the unit load is between S and M, the shearing force at S is $R-1$.

The influence line for shearing force is, therefore, the same as the influence line for R, when the unit load is between L and S, and shows a difference of 1 ton from the influence line for R when the unit load is to the right of S. (e)

Influence Line for Bending Moment at 6 ft. from M. So long as the unit load is to the left of S, the bending moment at S is $6R$ ft.-tons, and the influence line for bending moment at S is of the same shape as that for R. When the unit load is to the right of S, the bending moment at S is $6R-(6-a)$ ft.-tons.

From the ordinates representing $6R$, at any value of a between 0 and 6, a value of $(6-a)$ is subtracted to give the bending moment at S for this position of the unit influence load. (f)

Influence Line for Deflection at 6 ft. from M. (1) Using the method of Area Moments (see Chapter 2) find the expression for the upward displacement of S, due to R acting alone. (g) This is

$$\frac{1}{EI}(36R \times 4 + 108R \times 8)$$

$$= \frac{1008}{EI} R \ ft.$$

(2) Similarly, find the downward displacement of S due to the unit load acting at any section between S and M. (h)

$$-\frac{1}{EI}\left\{48(18-a)+24(6-a)\right\}$$

$$= -\frac{144}{EI}(7-9K) \ \text{where} \ K=\frac{a}{18}.$$

(3) Find the downward displacement of S due to the unit load acting at any section between L and S. (k)

$$-\frac{1}{EI}\left[\frac{1}{2}(18-a)^2\left\{12-\frac{1}{3}(18-a)\right\}\right]$$

$$= -\frac{972}{EI}\left\{(1-K)^2(1+K)\right\}.$$

8.1

The general expression for the final deflection at S (the equation to the required influence line) is found by taking the algebraic sum of the upward displacement at S due to R, and the downward displacement at S due to the unit load at any section distant a from M.

When the unit load is between S and M, the deflection at S is

$$\frac{1}{EI}\left\{1008R - 144(7-9K)\right\}.$$

When the unit load is between L and S, the deflection at S is

$$\frac{1}{EI}\left[1008R - 972\left\{(1-K)^2(1+K)\right\}\right]$$

The evaluation of these expressions by the tabular method gives the values of the influence line ordinates for selected positions of the unit load.

K	R	$1008R$	$(1-K)^2$	$(1+K)$	$(7-9K)$	Deflection at $S \times EI$
0	1·00	1008	—	—	7·00	0
0·2	0·70	706	—	—	5·20	−43
0·33	0·52	524	0·45	1·33	4·00	−52
0·4	0·43	433	0·36	1·40	—	−57
0·6	0·21	212	0·16	1·60	—	−37
0·8	0·06	60	0·04	1·80	—	−10
1·0	0	0	0	2·00	—	0

Every ordinate of this influence line represents the deflection at S when the unit load is at the section indicated by the ordinate.   (I)

Exercises

Draw *Influence Lines* for the beams in problems 2.1 and 3.1.

$\mathbf{8.2}$ This beam is statically indeterminate to the first degree, and thus only one influence line need be constructed by the method used in Problem **8.1**. Thereafter, the influence lines for other quantities can be drawn by the methods of statics.

As before, remove R_B and apply a unit load in its place. Then, by Macauley's Method for simple deflections, taking $a \geqslant 10$ ft.:

$$EI\frac{d^2y}{da^2}=\frac{12}{22}a-(a-10)$$

$$EI\frac{dy}{da}=\frac{6}{22}a^2-\frac{1}{2}(a-10)^2+A$$

$$EI\,y=\frac{a^3}{11}-\frac{1}{6}(a-10)^3+Aa+B.$$

When $a=0$, $y=0$ (omitting 2nd term), $B=0$.
When $a=22$, $y=0$, $A=-30\cdot9$.
Thus, when $a<10$ ft.

$$y=\frac{1}{EI}\left(\frac{a^3}{11}-30\cdot9a\right) \qquad \varDelta=-\frac{218}{EI}\ \text{(when }a=10\text{ ft.)}$$

and when $a>10$ ft.

$$y=\frac{1}{EI}\left\{\frac{a^3}{11}-\frac{1}{6}(a-10)^3-30\cdot9a\right\}.$$

As in Problem **8.1**

$$R_B=\frac{y}{\varDelta}=\frac{1}{218}\left(30\cdot9a-\frac{a^3}{11}\right)$$

when a is less than 10 ft.

$$R_B=\frac{y}{\varDelta}=\frac{1}{218}\left\{30\cdot9a-\frac{a^3}{11}+\frac{1}{6}(a-10)^3\right\}$$

when a is greater than 10 ft.

a ft.	$30\cdot9a$	$a^3/11$	$\frac{1}{6}(a-10)^3$	R_B tons
4	123·6	5·8	—	0·54
8	247·2	46·6	—	0·92
10	309·0	91·0	0	1·00
12	370·8	157·1	1·3	0·99
16	494·4	372·4	36·0	0·73
20	618·0	727·2	166·7	0·26
22	680·0	968·0	288·0	0

The value of R_B for any position of the unit load having been obtained, and the influence line drawn (c), it is relatively easy, by addition and subtraction of ordinates, to draw the influence lines for other important quantities.

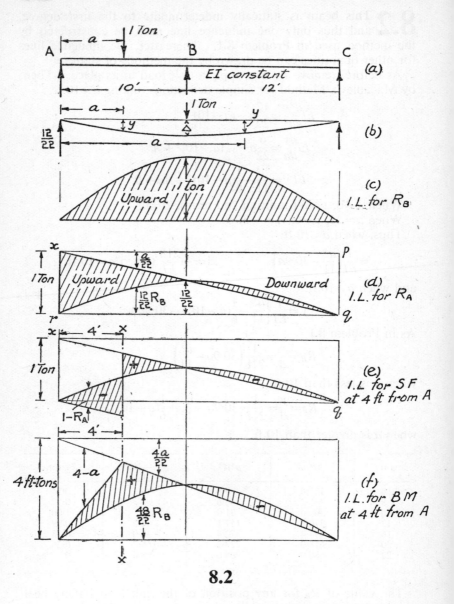

Influence Line for R_A. Taking moments about C, where bending moment is zero,

$$22R_A + 12R_B - (22-a) = 0.$$

$$R_A = -\frac{12}{22}R_B + 1 - \frac{a}{22}.$$

Draw the rectangle $xpqr$ having a vertical height of 1 ton. (**d**) At the bottom of this rectangle draw the R_B influence line, with each ordinate reduced to 12/22 of its value in (**c**). From the top of the rectangle draw ordinates of $a/22$ downwards (i.e. draw the diagonal).

Now put the point of a pencil on the curved line, anywhere along its length. Moving the pencil in the negative direction (downwards) through $-\frac{12}{22}R_B$ it reaches the base line rq. Continuing with a positive movement of unity, the pencil point reaches the line xp, and a negative movement of $-a/22$ gives a final point on the diagonal. Thus the ordinate between the curve and the diagonal represents the value of R_A. (**d**)

Influence Line for Shearing Force at 4 ft. from A. When the unit load is more than 4 ft. from A, the shearing force at section XX (**e**) is merely R_A and the influence line is a repetition of the previous one. Again, by covering the right-hand portion of the beam, from XX to C, it will be seen that when the unit load is between A and XX the shearing force at XX is $R_A - 1$, and all the ordinates of (**d**) are altered by unity, measured downwards from xq.

Influence Line for Bending Moment at 4 ft. from A. When the unit load is more than 4 ft. from A, the bending moment at the section XX (**f**) is merely $4R_A$ ft.-tons. The influence line for this portion of the beam is thus a repetition of (**c**) or (**d**) drawn to a different scale.

When the unit load is less than 4 ft. from A, the bending moment at A is

$$4R_A - (4-a) \text{ ft.-tons.}$$

The ordinate $(4-a)$ is drawn downwards from the diagonal.

8.3 In this beam there are two redundant reactions, R_B and R_C. When these are removed, the beam is in a statically determinate condition, and when the unit influence load is at some point on the beam, the deflections at B and C are δ_B and δ_C. This is the same beam as that in Problem **8.2**, but with one extra support.

As in Problem **8.2**, a unit load is placed at B and the elastic curve of the beam determined. In addition, a unit load must be placed at C, as a first step to finding the value of R_C. The elastic curve of the beam must also be found for this condition. **(c)** and **(d)**

The values of R_B and R_C will be found, in this problem, by the solution of two simultaneous equations. Each equation is formed by writing, in algebraic terms, that the upward deflection (at B or at C) caused by the reactions is numerically equal to the downward deflection due to the unit influence load at any point on the beam.

First Equation (Deflection at B). The upward deflection at B, due to the two redundant reactions, is composed of two portions: **(c)**

(1) $R_B \times$ the deflection at B due to a unit load at B (δ_{BB}).

(2) $R_C \times$ „   „ B „   „   ., C (δ_{BC}).

The values of R_B and R_C, then, are such that

$$R_B\delta_{BB} + R_C\delta_{BC} + \delta_B = 0.$$

Second Equation (Deflection at C). The upward deflection at C due to the two redundant reactions is composed of two portions: **(d)**

(1) $R_C \times$ the deflection at C due to a unit load at C (δ_{CC}).

(2) $R_B \times$ „   „ C „   „   „ B (δ_{CB}).

The second equation, made up in the same way as the first, is

$$R_B\delta_{CB} + R_C\delta_{CC} + \delta_C = 0.$$

It is now necessary to find, by the methods of statics, the various values of the deflections included in the two equations. This can be done by a consideration of the statically determinate beams shown in **(c)** and **(d)**.

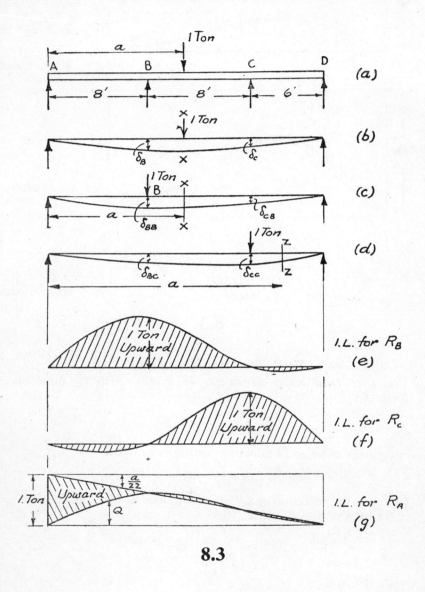

(a)

(b)

(c)

(d)

I.L. for R_B
(e)

I.L. for R_C
(f)

I.L. for R_A
(g)

8.3

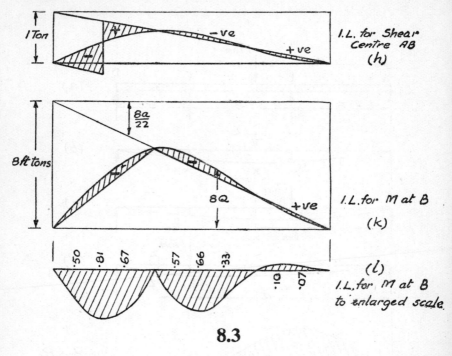

I.L. for Shear Centre AB
(h)

I.L. for M at B
(k)

(l)
I.L. for M at B
to enlarged scale

8.3

Elastic Curve of Simply Supported Beam of 22 ft. Span.

(a) *Unit Load acting downwards at B* (c). Taking moments about XX,

$$EI\frac{d^2\delta}{da^2}=\frac{14}{22}a-(a-8).$$

Integrating twice and evaluating constants as in **8.2**,

$$EI\delta_B=\tfrac{7}{66}a^3-\tfrac{1}{6}(a-8)^3-30{\cdot}55a.$$

This gives the deflection at XX with a unit load at B, or the deflection at B with a unit load at XX. (b)
When $a=8$ ft.

$$-\delta_B=+\delta_{BB}=+\frac{190{\cdot}1}{EI}\ ft.$$

When $a=16$ ft.

$$-\delta_B=+\delta_{CB}=+\frac{139{\cdot}6}{EI}\ ft.=\delta_{BC}\ \text{(by Maxwell's Theorem)}.$$

(b) *Unit Load acting downwards at C.* (d) Taking moments about ZZ,

$$EI\frac{d^2\delta}{da^2}=\frac{6}{22}a-(a-16).$$

Integrating twice and evaluating constants,

$$EI\delta_C=a^3/22-1/6(a-16)^3-20\cdot36a.$$

This gives the deflection at XX with a unit load at C, or the deflection at C with a unit load at XX. (b)
When a=8 ft.

$$-\delta_C=+\delta_{BC}=+\frac{139\cdot6}{EI}\ ft.$$

When a=16 ft.

$$-\delta_C=+\delta_{CC}=+\frac{139\cdot6}{EI}\ ft.$$

Substituting the values obtained above, in the first and second equations,

$$+190\cdot1R_B+139\cdot6R_C+0\cdot106a^3-\tfrac{1}{6}(a-8)^3-30\cdot55a=0 \left.\right\}$$
$$+139\cdot6R_B+139\cdot6R_C+0\cdot045a^3-\tfrac{1}{6}(a-16)^3-20\cdot36a=0 \left.\right\}$$

which, when solved, give

$$50\cdot5R_B=10\cdot19a-0\cdot061a^3+\tfrac{1}{6}(a-8)^3-\tfrac{1}{6}(a-16)^3.$$

This is the equation of the influence line for R_B. Remember, that in evaluating for the various values of a, the last two terms (according to Macauley's Method) are omitted if the term within the bracket is negative.

a ft.	10·19a	0·061a³	⅙(a−8)³	⅙(a−16)³	50·5R_B	R_B tons
2	20·4	0·5	—	—	19·9	+0·39
4	40·8	3·9	—	—	36·9	+0·73
6	61·1	13·1	—	—	48·0	+0·95
8	81·5	31·0	0	—	50·5	+1·00
10	101·9	60·6	1·3	—	42·6	+0·84
12	122·3	104·7	10·7	—	28·3	+0·57
14	142·7	166·3	36·0	—	12·4	+0·25
16	162·9	248·2	85·3	0	0	0
18	183·4	353·5	166·7	1·3	−4·7	−0·09
20	203·8	484·9	288·0	10·7	−3·8	−0·08

8.3

Influence Line for R_C. Similarly, solving the two equations for R_C

$$37{\cdot}1R_C = -20{\cdot}8a + 0{\cdot}033a^3 - 0{\cdot}122(a-8)^3 + \tfrac{1}{6}(a-16)^3$$

from which by a tabular calculation the values of the ordinates for the R_C influence line are found.

a ft.	2	4	6	8	10	12	14	16	18	20
R_C tons .	−0·11	−0·17	−0·15	0	+0·29	+0·63	+0·91	+1·00	+0·84	+0·48

Since the beam is statically indeterminate to the second degree it is necessary to find these two influence lines in this way. Thereafter the influence lines for other important quantities can be drawn by using statical methods.

Influence Line for R_A. Taking moments about D, where the bending moment is zero,

$$22R_A + 14R_B + 6R_C - (22-a) = 0.$$

$$R_A = -\left(\frac{14}{22}R_B + \frac{6}{22}R_C\right) + 1 - \frac{a}{22}.$$

This influence line may be plotted by the addition and subtraction of ordinates in the same way as for R_A in Problem **8.2**. This time, however, it is necessary to obtain, first of all, the combined curve for $Q = \frac{14}{22}R_B + \frac{6}{22}R_C$. This combined curve is then used as was the curve $\frac{12}{22}R_B$ in Problem **8.2**.

a ft.	2	4	6	8	10	12	14	16	18	20
$\frac{14}{22}R_B + \frac{6}{22}R_C$ . .	0·22	0·42	0·56	0·64	0·62	0·54	0·41	0·27	0·17	0·08

Influence Line for Shearing Force midway between A and B. The method used here is similar to that for (e) in Problem **8.2**, and will not be further described. (**8.3 (h)**)

Influence Line for Bending Moment at B. When the unit load is to the right of B, the bending moment at B is $8R_A$ ft.-tons. The influence line, from B to D, is thus of the same shape as that for R_A but is drawn to a different scale.

When the unit load is between A and B, the bending moment at B is

$$8R_A - (8-a).$$

Thus the "$8R_A$" ordinates between A and B must all be decreased by an amount $(8-a)$.

The shaded area (k) is the resultant shape of the influence line diagram for bending moment at B. Each ordinate represents the bending moment at B when the unit load is on the beam at a point directly above the ordinate in question.

Fig. (k) shows that if any accurate measurements are required from such influence lines, this method of combining ordinates may result in a scale which is too small. In such instances, the true values of each ordinate must be worked out, and the influence line plotted on a horizontal base to a larger scale, as is shown in (l).

Exercises

After studying the whole of this chapter, draw *Influence Lines* for the structures in problems 2.5, 2.7, 2.10, 3.4, 3.6, 4.1, 4.2, 4.3, 4.4, 5.1, 5.4, 6.3, 6.5, 6.7, 7.6.

8.4 This beam is statically indeterminate to the third degree, and influence lines must be drawn for three statically indeterminate quantities before the methods of statics can be employed.

The beam may be reduced to the statically determinate condition in several ways. For example, if the fixing moments at C and A and the reaction at A are removed, the beam is in the same statically determinate condition as the beams of Problems **8.2** and **8.3**.

In this solution of the problem, however, the statically indeterminate quantities have been chosen as the fixing moment at C (M_C), and the reactions at B and C (R_B and R_C). These are all assumed to act in a positive direction (see Fig. **2.02**). The signs resulting from the following calculations will then be the correct ones.

When the statically indeterminate quantities have been removed the beam becomes a cantilever, and a unit load at any section XX causes deflections at B and C equal to δ_B and δ_C respectively. The inclination of the beam at C under this loading is θ_C. **(c)**

Following the same procedure as in previous examples in this chapter, and utilising the principle of Maxwell's Theorem, we place a unit load at B, a unit load at C, and a unit moment at C. These, in order to avoid confusion, in a complex problem of this nature act in the positive direction. The results of such loading are shown in **(d)**, **(e)** and **(f)**.

If there is, in actual fact, no settlement or rotation of the supports, the effects of the unknown, statically indeterminate forces and moment must be equal and opposite to the effects of the unit influence load.

(1) The upward deflection ($+$) of B due to the action of the three statically indeterminate quantities is equal and opposite to the deflection of B ($-$) due to the unit load at any section XX.

(2) The upward deflection ($+$) of C due to the action of the three statically indeterminate quantities is equal and opposite to the deflection of C ($-$) due to the unit load at any section XX.

(3) The counter-clockwise rotation, from the horizontal, ($-$) of the tangent at C due to the action of the unit load at section XX is equal and opposite to the rotation at C due to the three statically indeterminate quantities.

When these three statements are translated into symbols three equations are obtained, from which general expressions for R_B, R_C and M_C may be calculated, and their influence lines drawn.

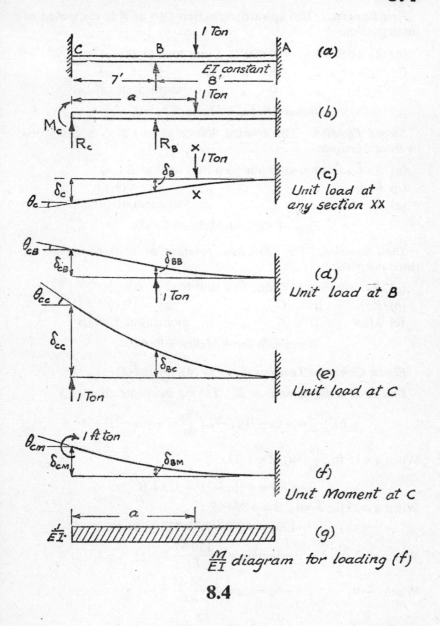

(a)

(b)

(c) Unit load at any section XX

(d) Unit load at B

(e) Unit load at C

(f) Unit Moment at C

(g) $\frac{M}{EI}$ diagram for loading (f)

8.4

211

8.4

First Equation. The upward deflection (+) at B is composed of three portions:

(a) $R_B \times$ deflection of B due to a unit load at $B(\partial_{BB})$
(b) $R_C \times$ „ B „   „   „ $C(\delta_{BC})$
(c) $M_C \times$ „ B „   „   moment at $C(\delta_{BM})$

$$R_B \delta_{BB} + R_C \delta_{BC} + M_C \delta_{BM} + \delta_B = 0.$$

Second Equation. The upward deflection (+) at C is composed of three portions:

(a) $R_B \times$ deflection at C due to a unit load at $B\,(_{CB})$
(b) $R_C \times$ „ C „   „   „ $C(\delta_{CC})$
(c) $M_C \times$ „ C „   „   moment at $C(\delta_{CM})$

$$R_B \delta_{CB} + R_C \delta_{CC} + M_C \delta_{CM} + \delta_C = 0.$$

Third Equation. The clockwise rotation at C is composed of three portions:

(a) $R_B \times$ rotation at C due to a unit load at $B(\theta_{CB})$
(b) $R_C \times$ „ C „   „   „ $C(\theta_{CC})$
(c) $M_C \times$ „ C „   „   moment at $C(\theta_{CM})$

$$R_B \theta_{CB} + R_C \theta_{CC} + M_C \theta_{CM} + \theta_C = 0.$$

Elastic Curves for Loadings shown in (d), (e) *and* (f).

Unit Load acting upwards at B. Taking moments about XX,

$$-EI\frac{d^2\delta_B}{da^2} = +(a-7) \; ; \quad -EI\frac{d\delta_B}{da} = +\tfrac{1}{2}(a-7)^2 + A$$

When $a = 15$ ft., $\dfrac{d\delta}{da} = 0$; $A = -32$.

$$-EI\delta_B = +\tfrac{1}{6}(a-7)^3 - 32a + B.$$

When $a = 15$ ft., $\delta = 0$; $B = +394 \cdot 67$

$$-EI\delta_B = +\tfrac{1}{6}(a-7)^3 - 32a + 394 \cdot 67.$$

When $a = 7$ ft. $-\delta_B = \delta_{BB} = +\dfrac{170 \cdot 67}{EI}$ *ft.*

When $a - 0$ $-\delta_B = \delta_{CB} = +\dfrac{394 \cdot 67}{EI}$ *ft.*

When $a = 0$ $-\dfrac{d\delta}{da} = \theta_{CB} = -\dfrac{32}{EI}$ *rad.*

Unit Load acting upwards at C. The elastic curve for a cantilever of length l and carrying a unit load at the outer end was determined in Problem **8.1.**

$$-EI\delta_C = \tfrac{1}{6}a^3 - 112\cdot5a + 1125.$$

When $a=0$ $\qquad -\delta_C = \delta_{CC} = +\dfrac{1125}{EI}$ ft.

When $a=7$ ft. $\qquad -\delta_C = \delta_{BC} = +\dfrac{394\cdot67}{EI}$ ft.

When $a=0$ $\qquad -\dfrac{d\delta_C}{da} = \theta_{CC} = -\dfrac{112\cdot5}{EI}$ rad.

Unit Moment acting clockwise at C. Using the method of Area Moments and the M/EI diagram shown in (**g**) the following values are found:

When $a=7$ ft. $\qquad \delta_{BM} = +\dfrac{32}{EI}.$

When $a=0$ $\qquad \delta_{CM} = +\dfrac{112\cdot5}{EI}.$

When $a=0$ $\qquad \theta_{CM} = -\dfrac{15}{EI}.$

From the M/EI diagram for the loading shown in (**c**), the slope at C due to a unit load at any section XX of the beam is

$$\theta_C = +\tfrac{1}{2}a^2 - 15\cdot0a + 112\cdot5.$$

Substituting the values obtained in this section in the first, second and third equations, the following numerical relationships result:

(1) $+170\cdot7R_B + 394\cdot7R_C + 32\cdot0M_C - \tfrac{1}{6}(a-7)^3 + 32\cdot0a - 394\cdot7 = 0.$
(2) $+394\cdot7R_B + 1125\cdot0R_C + 112\cdot5M_C - \tfrac{1}{6}a^3 + 112\cdot5a - 1125\cdot0 = 0.$
(3) $-32\cdot0R_B - 112\cdot5R_C - 15\cdot0M_C + \tfrac{1}{2}a^2 - 15\cdot0a + 112\cdot5 = 0.$

The solution of these three simultaneous equations is tedious, but perfectly straightforward. It requires, however, to be carried out to as many significant figures as is possible with the means available, if errors due to approximation are not to result.

The final solutions are:

$$R_B = +0\cdot0096(a-7)^3 - 0\cdot0053a^3 + 0\cdot0573a^2.$$
$$R_C = -0\cdot0053(a-7)^3 + 0\cdot0035a^3 - 0\cdot0449a^2 + 1\cdot0.$$
$$M_C = +0\cdot0191(a-7)^3 - 0\cdot0150a^3 + 0\cdot2476a^2 - a.$$

213

8.4

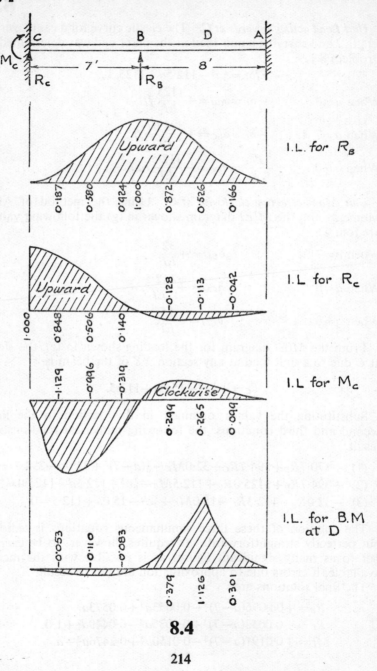

C, B, D, A

M_c

R_c 7' R_B 8'

I.L. for R_B

Upward

0·187 0·580 0·824 1·000 0·872 0·526 0·166

I.L for R_c

Upward

1·000 0·848 0·506 0·140 −0·128 −0·113 −0·042

I.L for M_c

−1·129 −0·996 −0·319 Clockwise 0·299 0·265 0·099

I.L. for B.M at D

−0·053 −0·110 −0·083 0·379 1·126 0·301

8.4

214

These equations represent the required influence lines, whose ordinates are now evaluated for selected values of a. As before, if the quantity inside the bracket is negative, that term does not appear in the evaluation. In other words, the $(a-7)^3$ terms are evaluated only for sections where a is greater than 7 ft.

The three influence lines are shown on page **214**, and from these other influence lines may be built up by the methods of statics, in the way which was demonstrated in previous problems of this chapter.

Influence Line for Bending Moment at the Centre of BA. When the unit load is to the right of D the bending moment at D is

$$M_D = M_C + 11 R_C + 4 R_B.$$

When the unit load is to the left of D the bending moment at D is

$$M_D = M_C + 11 R_C + 4 R_B - (11 - a).$$

Using the values of M_C, R_B and R_C for each selected value of a, the ordinates of the required influence line can be obtained.

INFLUENCE LINES FOR TWO-HINGED AND FIXED PORTALS

8.5 A two-hinged portal frame is statically indeterminate to the first degree, and the influence line for H_A must first be drawn. Using the arguments advanced in the introduction to this chapter (**8.0**), it can be shown that the effect of a unit load acting at the hinge A, and in the direction of the horizontal reaction H_A, can be used to obtain, directly, the influence line for H_A.

With such a load acting (**a**), the displacement of any point on the frame (horizontally for the columns and vertically for the beam) divided by the horizontal displacement of A, gives the value of the influence line ordinate for that section. In other words, as for the beams considered above, the deflected shape of the frame when a unit load acts in the line of action of the statically indeterminate reaction is of the same form as the influence line for that reaction.

The force H_A acting on A calls into play an equal reaction (H) at D. The hinge D remains stationary, and the frame deflects from its original (dotted) position, as shown in (**a**). However, in order to facilitate the calculation of the magnitude of d, we may assume both A and D to move, and use Fig. (**b**) in the preliminary calculations.

215

8.5

Vertical Displacement of a Point on Beam BC. Taking moments about XX,

$$EI\frac{d^2\delta}{dx^2} = -15; \quad EI\frac{d\delta}{dx} = -15x + A.$$

From symmetry, when $x = 5$ ft., $d\delta/dx = 0$; $A = +75$

$$EI\delta = -15\frac{x^2}{2} + 75x + B.$$

When $x = 0$, $\delta = 0$, $B = 0$

$$EI\delta = -15\frac{x^2}{2} + 75x.$$

Also $\quad EI\frac{d\delta}{dx}$ (for $x=0$) $= +75$ (slope at B).

Horizontal Displacement of a point on Column AB. Taking moments about YY,

$$EI\frac{d^2\delta}{da^2} = -a; \quad EI\frac{d\delta}{da} = -\frac{a^2}{2} + C.$$

When $a = 15$ ft., $\frac{d\delta}{da} =$ slope at $B = +\frac{75}{EI}$ $\therefore$ $C = +187\cdot5$.

$$EI\delta = -\frac{a^3}{6} + 187\cdot5a + D.$$

When $a = 15$ ft., $\delta = 0$ $\therefore D = -2250$

$$EI\delta = -\frac{a^3}{6} + 187\cdot5a - 2250.$$

The displacements which represent the influence line for H_A are those which occurred from the original position (**a**). So far, only the value of δ from Fig. (**b**) has been obtained.

For the left column, the required displacements (**a**) are $(\delta + d)$, and for the right column, the displacement from the original position is $(d - \delta)$. By using the following tabular form of calculation, the displacements required can be calculated and the influence line for H_A drawn. The value of d is $-2250/EI$ ft.

In Fig. (**a**) it can be seen that the displacements of points on the frame from their original positions are negative (inward) for the left column and positive (upward or outward) for the right column and beam. These displacements, divided by $\Delta = 2d = 4500/EI$, give the values of the influence line ordinates for H_A.

216

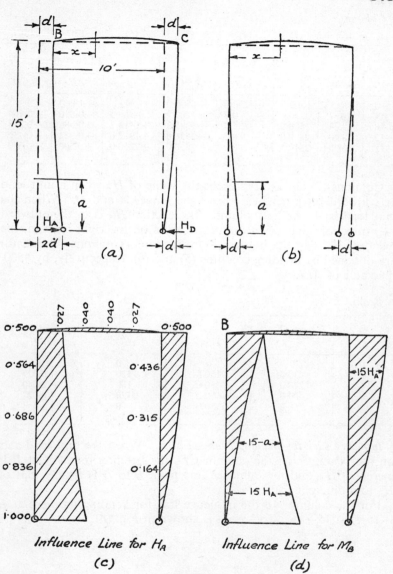

Influence Line for H_A

(c)

Influence Line for M_B

(d)

8.5

Columns.

(1)	(2)	(3)	(4)	(5)	(6)	(7)	(8)
						I.L. Ordinates for H_A	
a	$a^3/6$	$187 \cdot 5a$	$EI\delta$	$EI(\delta+d)$	$-EI(d-\delta)$	AB	CD
0	0	0	$-2250 \cdot 0$	$-4500 \cdot 0$	0	$+1 \cdot 000$	0
4	$10 \cdot 7$	$750 \cdot 0$	$-1510 \cdot 7$	$-3760 \cdot 7$	$+739 \cdot 3$	$+0 \cdot 836$	$-0 \cdot 164$
8	$85 \cdot 3$	$1500 \cdot 0$	$-835 \cdot 3$	$-3085 \cdot 3$	$+1414 \cdot 7$	$+0 \cdot 686$	$-0 \cdot 315$
12	$288 \cdot 0$	$2250 \cdot 0$	$-288 \cdot 0$	$-2538 \cdot 0$	$+1962 \cdot 0$	$+0 \cdot 564$	$-0 \cdot 436$
15	$562 \cdot 5$	$2812 \cdot 5$	0	$-2250 \cdot 0$	$+2250 \cdot 0$	$+0 \cdot 500$	$-0 \cdot 500$

Columns (7) and (8) represent the value of H_A when a unit load acts inwards at a height a above the hinges A and D. When the unit load is on the left column, the reaction H_A is in the outward (+) direction, and when the unit load is on the right column, the reaction H_A is in the inward (−) direction. Columns (7) and (8) are obtained by dividing columns (5) and (6), respectively, by 4500, the value of ΔEI.

Beam.

x	$15\dfrac{x^2}{2}$	$75x$	$EI\delta$	I.L. Ordinate for H_A
0	0	0	0	0
2	$30 \cdot 0$	$150 \cdot 0$	120	$-0 \cdot 027$
6	$270 \cdot 0$	$450 \cdot 0$	180	$-0 \cdot 040$
8	$480 \cdot 0$	$600 \cdot 0$	120	$-0 \cdot 027$
10	$750 \cdot 0$	$750 \cdot 0$	0	0

Influence Line for Bending Moment at B. When the unit load acts on the beam, or on the column CD, the bending moment at B is merely $15H_A$, and the influence line from B to D is merely that of H_A to a different scale.

For the column AB the influence line for bending moment at B is $15H_A-(15-a)$. The result is shown in Fig. (d).

8.6 As in Problem **8.4**, the three statically indeterminate quantities H_A, V_A, and M_A are assumed to act in the positive direction. This is clockwise for M_A, outward for H_A and upward for V_A. As before, the statically indeterminate forces and moments are removed, and the frame considered in the statically determinate condition.

Since, in actual fact, there is no vertical or horizontal movement of A, and no rotation of the tangent at A, the effect of the unit influence load in causing such displacements at A (**b**) must be exactly counterbalanced by the effects of H_A, V_A, and M_A. The effects of unit forces and moments are shown separately in (**c**), (**d**), and (**e**).

As in Problem **8.4**, therefore, we may write three equations, each one dealing with one of the three displacements, and from these three equations the values of H_A, V_A and M_A may be determined.

First Equation. Rotation of the tangent at A due to statically indeterminate forces and moment, plus the rotation of the tangent at A due to the unit influence load, equals zero.

$$H_A\theta_H + V_A\theta_V + M_A\theta_M + \theta_L = 0.$$

Second Equation. Vertical displacement of A due to H_A, V_{AA} and M_A, plus vertical displacement of A due to unit influence load, equals zero.

$$H_A d_{VH} + V_A d_{VV} + M_A d_{VM} + d_{VL} = 0.$$

Third Equation. Horizontal displacement of A due to H_A, V_A and M_A, plus horizontal displacement of A due to unit influence load, equals zero.

$$H_A d_{HH} + V_A d_{HV} + M_A d_{HM} + d_{HL} = 0.$$

Determination of Numerical Values. Methods of determining the values of the displacements at A might be similar to those employed in previous problems. Since, however, there is a permanently vertical tangent at D, a very simple solution can be obtained by the use of the Method of Area Moments (see Chapter 2).

The rotation at A relative to the tangent at D is the sum of the M/EI areas between A and D.

The vertical displacement of A (i.e. parallel to the tangent at D) is the moment of the M/EI areas about a vertical line through A.

The horizontal displacement of A (i.e. at right angles to the tangent at D) is the moment of the M/EI areas about a horizontal line through A.

The M/EI areas for unit forces and moment at A are shown hatched in (**c**), (**d**) and (**e**).

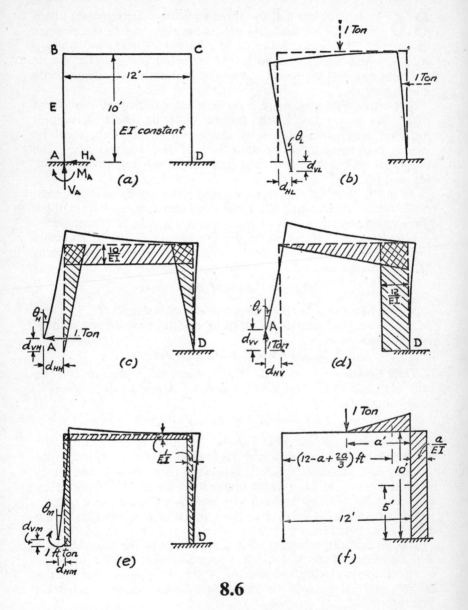

Unit Load acting Horizontally at A (Fig. (c)).
Area of M/EI diagram between D and A

$$=\frac{1}{EI}(50+120+50)=\frac{220}{EI}=\theta_H.$$

Moment of the M/EI diagram between D and A, about a vertical through A

$$=\frac{1}{EI}(120\times 6+50\times 12)=\frac{1320}{EI}=d_{VH}.$$

Moment of the M/EI diagram between A and D about a horizontal through A

$$=\frac{1}{EI}(120\times 10+2\times 50\times 6\cdot 67)=\frac{1866\cdot 7}{EI}=d_{HH}.$$

Unit Load acting Vertically at A (Fig. (d)).
Area of M/EI diagram between A and D

$$=\frac{1}{EI}(72+120)=\frac{192}{EI}=\theta_V.$$

Moment of the M/EI diagram between D and A, about a vertical through A

$$=\frac{1}{EI}(72\times 8+120\times 12)=\frac{2016}{EI}=d_{VV}.$$

Moment of the M/EI diagram between D and A, about a horizontal through A

$$=\frac{1}{EI}(72\times 10+120\times 5)=\frac{1320}{EI}=d_{HV}.$$

Unit Moment acting clockwise at A (Fig. (e)).
Area of M/EI diagram between D and A

$$=\frac{1}{EI}(2\times 12+10)=\frac{32}{EI}=\theta_M.$$

Moment of the M/EI diagram between D and A, about a vertical through A

$$=\frac{1}{EI}(12\times 6+10\times 12)=\frac{192}{EI}=d_{VM}.$$

Moment of the M/EI diagram between D and A about a horizontal through A

$$=\frac{1}{EI}(2\times 10\times S+12\times 10)=\frac{220}{EI}=d_{HM}.$$

221

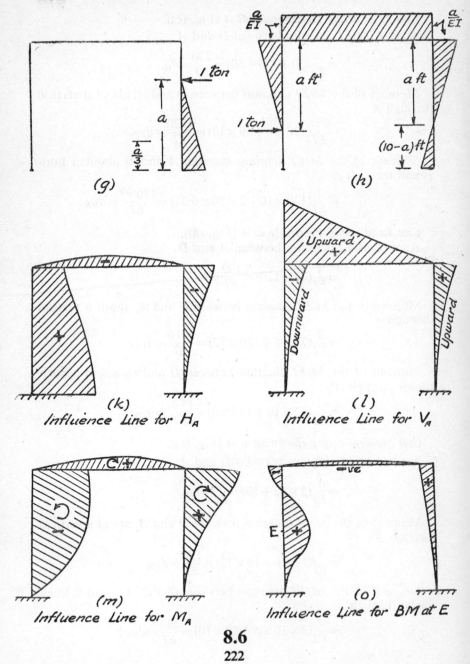

8.6

$\frac{a}{EI}$

$\frac{a}{EI}$

1 ton

a

$\frac{a}{3}$

a ft

a ft

1 ton

(10-a) ft

(g)

(h)

Upward

Downward

Upward

(k)
Influence Line for H_A

(l)
Influence Line for V_A

(m)
Influence Line for M_A

E

(o)
Influence Line for BM at E

Substituting these values in the first, second and third equations, we have the following relationships :

$$220H_A + 192V_A + 32M_A + \theta_L \times EI = 0$$
$$1320H_A + 2016V_A + 192M_A + d_{VL} \times EI = 0$$
$$1867H_A + 1320V_A + 220M_A + d_{HL} \times EI = 0$$

The simultaneous solution of these equations, although tedious, is a straightforward operation. It must, however, be carried out to a considerable degree of accuracy if reliable results are to be expected. The numerous differences which must be obtained between quantities of the same order may cause a wide divergence from the true values.

The student is advised to try a solution of the general case, using symbols for the height and span of the frame, and to compare the influence lines obtained with those given below.

Solution of the Equations. The expressions for H_A, V_A, and M_A, obtained by solving the three equations, are as follows:

$$H_A = EI(+0.01941\theta_L - 0.00282d_{HL}).$$
$$V_A = EI(+0.00694\theta_L - 0.00116d_{VL}).$$
$$M_A = EI(-0.20636\theta_L + 0.00694d_{VL} + 0.01941d_{HL}).$$

Determination of θ_L, d_{VL} and d_{HL}. Before the influence lines for the three statically indeterminate quantities can be drawn, the numerical values of θ_L, d_{VL} and d_{HL} must be found. These values will differ according as the unit influence load acts on the columns of the frame or on the beam, and thus three separate evaluations must be made.

Figs. (f), (g) and (h) show the unit influence load in the three possible positions and the M/EI diagrams which govern the magnitude of the displacements shown in Fig. (b).

Unit Load on the Beam BC (Fig. (f)). Slope at $A(\theta_L)$ equals the area of the M/EI diagram between D and A

$$= \frac{1}{EI}(-0.5a^2 - 10a) = \theta_L.$$

Vertical displacement of $A(d_{VL})$ equals the moment of the area of the M/EI diagram between D and A about a vertical through A

$$= \frac{1}{EI}\left(-0.5a^2\left(12 - a + \frac{2a}{3}\right) - 10a \times 12\right) = \frac{1}{EI}\left(\frac{a^3}{6} - 6a^2 - 120a\right) = d_{VL}.$$

Horizontal displacement of A (d_{HL}) equals the moment of the area of the M/EI diagram between D and A about a horizontal through A

$$= \frac{1}{EI}(-0.5a^2 \times 10 - 10a \times 5) = \frac{1}{EI}(-5a^2 - 50a) = d_{HL}.$$

8.6

Unit Load on the Column CD (Fig. (g)).   Slope at A equals area of M/EI diagram between D and A

$$=\frac{1}{EI}(-0{\cdot}5a^2)=\theta_L.$$

Vertical displacement of A equals moment of M/EI diagram between D and A about vertical through A

$$=\frac{1}{EI}(-0{\cdot}5a^2\times 12)=\frac{1}{EI}(-6a^2)=d_{VL}.$$

Horizontal displacement of A equals moment of M/EI diagram between D and A about horizontal through A

$$=\frac{1}{EI}\left(-0{\cdot}5a^2\times\frac{a}{3}\right)=\frac{1}{EI}\left(-\frac{a^3}{6}\right)=d_{HL}.$$

Unit Load on Column AB (Fig. (h)).   Slope at A equals the area of M/EI diagram between D and A

$$=\frac{1}{EI}\{-0{\cdot}5a^2-12a-0{\cdot}5a^2+0{\cdot}5(10-a)^2\}$$

$$=\frac{1}{EI}(-0{\cdot}5a^2-22a+50)=\theta_L.$$

Vertical displacement of A equals moment of M/EI diagram between D and A about a vertical through A

$$=\frac{1}{EI}\{-0{\cdot}5a^2\times 12-12a\times 6+0{\cdot}5(10-a)^2\times 12\}$$

$$=\frac{1}{EI}(-192a+600)=d_{VL}.$$

Horizontal displacement of A equals moment of M/EI diagram between D and A about a horizontal through A

$$=\frac{1}{EI}\left\{-a^2\left(10-a+\frac{2a}{3}\right)-120a+\frac{(10-a)^3}{6}\right\}$$

$$=\frac{1}{EI}\left(+\frac{a^3}{6}-5a^2-170a+166{\cdot}67\right)=d_{HL}.$$

Substituting the values of θ_L, d_{HL} and d_{VL} just obtained, in the general expressions for H_A, V_A and M_A, we find the equations of the influence lines as follows:

Influence Lines for H_A, V_A and M_A. (1) When the unit influence load is acting horizontally on the column AB

$$H_A = -0\cdot0005a^3 + 0\cdot0044a^2 + 0\cdot0529a + 0\cdot50$$
$$V_A = \qquad\qquad -0\cdot0035a^2 + 0\cdot0694a - 0\cdot3472$$
$$M_A = +0\cdot0032a^3 + 0\cdot0061a^2 - 0\cdot0931a - 2\cdot9164.$$

Positive sign indicates an outward or upward force and a clockwise moment at A.

(2) When the unit influence load is acting vertically on the beam BC

$$H_A = \qquad\qquad +0\cdot0044a^2 - 0\cdot0529a$$
$$V_A = -0\cdot0002a^3 + 0\cdot0035a^2 + 0\cdot0694a$$
$$M_A = +0\cdot0012a^3 - 0\cdot0355a^2 + 0\cdot2598a.$$

Positive sign indicates an outward or upward force and a clockwise moment at A.

(3) When the unit influence load is acting horizontally on the column CD

$$H_A = +0\cdot0005a^3 - 0\cdot0097a^2$$
$$V_A = \qquad\qquad +0\cdot0035a^2$$
$$M_A = 0\cdot0032a^3 + 0\cdot0615a^2.$$

Positive sign indicates an outward or upward force and a clockwise moment at A.

It now remains to substitute values of a in these equations and so obtain the various influence line ordinates used in drawings Figs. (k), (l), and (m).

Influence Line for Bending Moment at E. As in previous problems, it is now possible to use the methods of statics to determine the influence line ordinates for other important quantities.

When the unit influence load is on CD, BC, or BE, the value of the bending moment at E is

$$M_A + 5H_A.$$

When the unit influence load is below E (acting horizontally), the value of the bending moment at E is

$$M_A + 5H_A - x$$

where x represents the distance between the unit load and E. Remembering to use the correct signs, we may now evaluate the influence line ordinates and draw Fig. (o).

15   225

INFLUENCE LINES FOR TWO-HINGED AND FIXED ARCHES

8.7 Influence lines for two-hinged and fixed arches are very important tools in the design of such structures, and the examples below show how they may be drawn for simple arch ribs. For arch ribs of more complex form the procedure may be more complex, but is based on the same principles.

The moment of inertia of this arch rib is assumed to vary as the secant of its slope.

As for portal frames, the first influence line which must be drawn is that for H. It is, therefore, necessary to obtain a general expression for H in terms of the variable a. The value of this expression will then be plotted as ordinates under successive positions of the unit load, to form the required influence line.

$$H = \frac{\displaystyle\int_0^{12} My\,dx}{\displaystyle\int_0^{12} y^2\,dx}$$

$$y = Cx(l-x) = \frac{x}{9}(l-x) \quad \text{(see Chapter 4).}$$

When x is less than a, and measured from the left abutment,

$$M = V_L x = \frac{l-a}{l}x.$$

When x is less than $l-a$ and is measured from the right abutment,

$$M = V_R x = \frac{a}{l}x.$$

Integrating over the whole span,

Numerator.

$$\int_0^{12} My\,dx = \int_0^a \frac{l-a}{l}\cdot\frac{x^2}{9}(12-x)dx + \int_0^{l-a} \frac{a}{l}\cdot\frac{x^2}{9}(12-x)dx$$

$$= \frac{a(12-a)}{108}(144+12a-a^2)\ lb.\text{-}in.^3$$

226

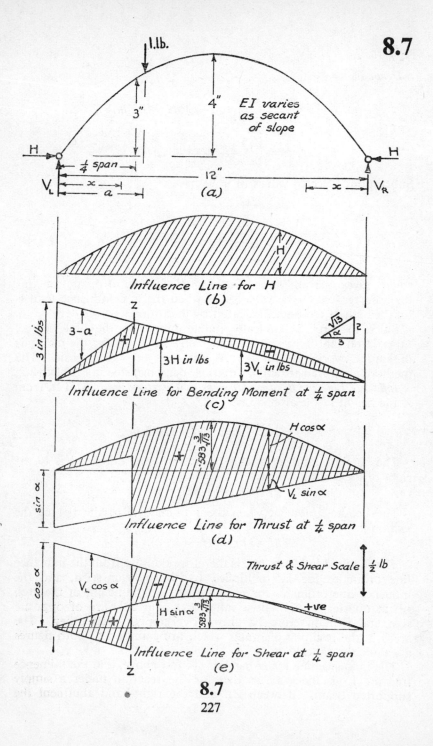

1.lb.

3"

4" *EI varies as secant of slope*

H ← H ⟶

$\frac{1}{4}$ *span*

12"

x

a

x

V_L V_R

(a)

Influence Line for H
(b)

3 in lbs

3 − a

Z

3H *in lbs* $3V_L$ *in lbs*

√13 / α / 3   2

Influence Line for Bending Moment at $\frac{1}{4}$ span
(c)

sin α

$\cdot 583 \frac{3}{\sqrt{13}}$

H cos α

V_L sin α

Influence Line for Thrust at $\frac{1}{4}$ span
(d)

cos α

V_L cos α

Thrust & Shear Scale $\frac{1}{2}$ lb

H sin α $\cdot 583 \frac{3}{\sqrt{13}}$

+ve

Z

Influence Line for Shear at $\frac{1}{4}$ span
(e)

8.7

8.7

Denominator.

$$\int_0^{12} y^2 dx = \int_0^{12} \frac{x^2}{81}(12-x)^2 dx = 103 \ in.^3$$

$$\therefore H = \frac{a(12-a)}{108 \times 103}(144 + 12a - a^2).$$

Substituting selected values of a we have

a	0	1	2	3	4	5	6
H	0	0·150	0·295	0·416	0·505	0·562	0·583

which gives the influence line for H. Every ordinate of this diagram represents the value of H when the unit influence load is on the arch at the point indicated by the ordinate.

Since the arch is statically indeterminate to the first degree, other important influence lines may now be drawn by the methods of statics, working with the H influence line as a basis. The methods employed are similar to those demonstrated in Problem **8.2**.

Influence Line for Bending Moment at Quarter-span. At 3 ft. from A the height of the arch rib is

$$y = \frac{x}{9}(12-x) = \frac{1}{3}(12-3) = 3 \ in.$$

The bending moment at this point when the unit load is to the right of the section is

$$M = 3V_L - 3H.$$

The bending moment when the unit load is to the left of the section is

$$M = 3V_L - 3H - (3-a).$$

The required influence line is therefore built up from the influence line ordinates for V_L multiplied by the quarter span, and the influence line ordinates for H multiplied by the height of the arch rib at quarter span. These values ($3V$ and $3H$) are of opposite sign and will therefore be drawn overlapping each other. The areas in the resulting diagrams which are common to both figures are mutually cancelling.

The influence line for H has been found above, and the influence line for V_L is the same as that for the reaction under a simply supported beam. Between ZZ and the right-hand abutment the

228

difference of the ordinates $3H$ and $3V_L$ is the required influence line ordinate, but between ZZ and the left-hand abutment a further reduction of $(3-a)$ must be made, as is shown in Fig. (c).

Influence Line for Thrust at Quarter-span. At this point the student is recommended to read Problem **4.5**, where the method of determining thrust and shear for a stationary loading is discussed.

As in Problem **4.5**, the first step is to determine the slope of the arch rib at the section in question. At quarter span ($x=3$ in.)

$$\frac{dy}{dx}=\frac{4}{3}-\frac{2x}{9}=\frac{4}{3}-\frac{6}{9}=\frac{2}{3},$$

whence **(c)**
$$\cos \alpha = \frac{3}{\sqrt{13}}$$

and
$$\sin \alpha = \frac{2}{\sqrt{13}}.$$

The thrust in the arch rib (compression positive) at quarter span when the unit load is to the right of ZZ, is

$$H \cos \alpha + V_L \sin \alpha.$$

The thrust when the unit load is to the left of quarter span (as in Fig. **4.5**) is given by the heavy lines as

$$H \cos \alpha + V_L \sin \alpha - \sin \alpha.$$

This time the basic influence lines have the same sign, and ordinates must be summed. Each ordinate of the H influence line is multiplied by the constant $\cos \alpha$, found above, and each ordinate of the V_L influence line by the constant $\sin \alpha$.

Between the left abutment and quarter span the sum of the $H \cos \alpha$ and $V_L \sin \alpha$ ordinates must be decreased by a constant value of $\sin \alpha$, which results in the influence line shown in Fig. **8.7 (d)**.

Influence Line for Shear at Quarter-span. When the unit load is to the right of quarter span, the shear at section ZZ is

$$V_L \cos \alpha - H \sin \alpha,$$

and when the unit load is to the left of quarter span, the shear at quarter span (ZZ) is

$$V_L \cos \alpha - H \sin \alpha - \cos \alpha.$$

Adding and subtracting ordinates graphically as in the sketching of previous influence lines, we have the influence line for shear at quarter span shown at **(e)**.

8.8 Before studying this problem the student should read and solve Problem **4.9**. In that problem the values of H, V_o, and M_o are obtained for a single system of loads. In the present problem the influence lines for these three unknown quantities are required. This demands several solutions, one for each chosen position of the unit load.

The first step is to determine the constant values which refer to the shape of the arch; the height of the rib above the springing line, the moment of inertia of the section, etc. As in all design of statically indeterminate structures, the sizes and proportions of the various parts of the structure must first be determined by a preliminary analysis before the bending moments, thrusts, shears and fibre stresses may be calculated. It is assumed that this preliminary analysis has been carried out and that the arch rib which it is proposed to investigate is of reinforced concrete, varying in thickness from 3 ft. at the springing to 1 ft. 6 in. at the crown. It is fixed at both abutments, and a 1 ft. breadth of the arch will be considered.

In order to illustrate clearly the method of obtaining influence lines for a "practical" arch of this kind, two " short cuts" have been used. *First*, it has been assumed that the student knows how to find the moment of inertia of a reinforced concrete section reinforced on both faces, and the final values of I only have been given. *Secondly*, in order to simplify the figures in the tables below, the arch span has been divided into a number of equal segments. In fact, the arch rib centre line should have been so divided. The assumption that Δx is equal to Δs (**a**) is not strictly correct, but the justification for its use here lies in the fact that the calculations are thus rendered less involved. In practice a large-scale drawing of the rib, divided along its centre line into a number of equal parts, could be used for the determination of x and y by direct scaling, each measurement being taken to the centre of the segment in question.

The first table, in which the "arch constants" are calculated, must first be thoroughly understood. The equation to the shape of the arch (which is parabolic in this instance) is used to determine x and h, and with the values of the moment of inertia of the cross-section these figures can be used to find the values in the other columns.

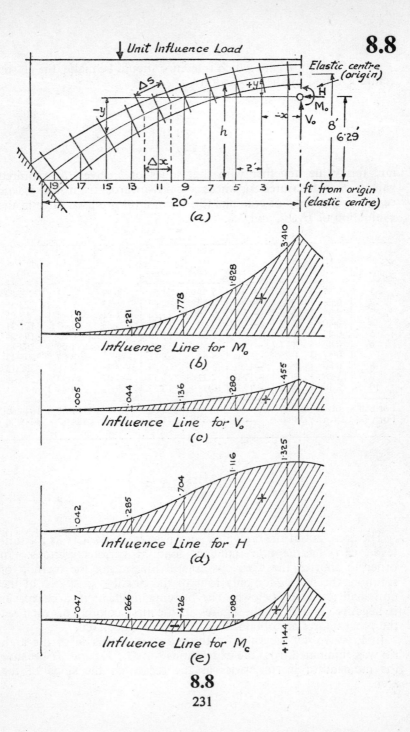

Unit Influence Load

Elastic centre (origin)

Influence Line for M_o
(b)

Influence Line for V_o
(c)

Influence Line for H
(d)

Influence Line for M_c
(e)

8.8

231

8.8

As in Problem **4.9**, to which reference should be made, the elastic centre is found by the expression

$$Z = \frac{\sum \frac{h}{I}}{\sum \frac{1}{I}}$$

and this value, together with the values of h, gives the required values of y measured upwards or downwards from the elastic centre. The sums of columns (2), (7) and (8) are used later in the evaluation of H, V_o, and M_o.

Section	I (1)	$1/I$ (2)	x (3)	h (4)	h/I (5)	y (6)	y^2/I (7)	x^2/I (8)
1	0·37	2·70	−1·0	7·98	21·57	1·69	7·72	2·70
2	0·38	2·62	−3·0	7·82	20·47	1·53	6·13	23·56
3	0·39	2·56	−5·0	7·50	19·18	1·21	3·75	63·95
4	0·40	2·49	−7·0	7·02	17·46	0·73	1·33	121·91
5	0·42	2·40	−9·0	6·38	15·30	0·09	0·02	194·24
6	0·46	2·17	−11·0	5·58	12·10	−0·71	1·09	262·45
7	0·57	1·75	−13·0	4·62	8·11	−1·67	4·89	296·43
8	0·80	1·25	−15·0	3·50	4·39	−2·79	9·76	282·15
9	1·28	0·78	−17·0	2·22	1·73	−4·07	12·92	225·42
10	2·10	0·48	−19·0	0·78	0·37	−5·51	14·45	171·84
For half span:	19·20	—	—	120·68	—	62·06	1644·65
For whole span:	38·40	—	—	241·36	—	124·12	3289·30

$$Z = \frac{\sum \frac{h}{I}}{\sum \frac{1}{I}} = \frac{241 \cdot 36}{38 \cdot 40} = 6 \cdot 29 \; ft.$$

The arch, cut at the crown as shown in (a), now acts as a cantilever, carrying the unit influence load at various sections. In order to shorten the work, while still illustrating the method of solution, the figures for only four of the possible positions of the unit load are given below. The bending moment in a cantilever is negative, and from the figures in the first column of each set the other figures are obtained and the sums calculated.

In Problem **4.9** the value of the moment of inertia of the arch rib was eliminated from the expressions for M_o, V_o, and H because the moment of inertia varies as the secant of the slope of the arch.

232

Section	Influence Load at Section (1)				Influence Load at Section (3)			
	M	M/I	Mx/I	My/I	M	M/I	Mx/I	My/I
1	0	—	—	—	0	—	—	—
2	−2	− 5·26	15·79	− 8·05	0	—	—	—
3	−4	−10·23	51·15	−12·38	0	—	—	—
4	−6	−14·93	104·48	−10·90	−2	−4·97	34·82	−3·63
5	−8	−19·19	172·66	− 1·73	−4	−9·59	86·33	−0·86
6	−10	−21·69	238·61	15·40	−6	−13·01	143·16	9·24
7	−12	−21·05	273·69	35·16	−8	−14·04	182·46	23·44
8	−14	−17·56	263·48	49·01	−10	−12·55	188·21	35·01
9	−16	−12·48	212·16	50·79	−12	− 9·36	159·12	38·10
10	−18	− 8·56	162·62	47·16	−14	− 6·66	126·48	37·21
For whole span:	−130·95	1494·64	164·46		—	−70·18	920·58	138·51

Section	Influence Load at Section (5)				Influence Load at Section (7)			
	M	M/I	Mx/I	My/I	M	M/I	Mx/I	My/I
6	−2	−4·34	47·72	3·08	0	—	—	—
7	−4	−7·02	91·23	11·72	0	—	—	—
8	−6	−7·53	112·92	21·00	−2	−2·51	37·63	7·00
9	−8	−6·24	106·08	25·40	−4	−3·12	53·04	12·70
10	−10	−4·75	90·35	26·20	−6	−2·85	54·21	15·72
For whole span:	−29·88	448·30	87·40		—	−8·48	144·88	35·42

In this problem, the expressions for M_o, V_o, and H are similar to those used in Problem 4.9, but the "area" used in the numerators of the three expressions is the area of the M/I diagram instead of the area of the M diagram. The expressions for the statically indeterminate quantities are

$$M_o = - \frac{\text{Area of } M/I \text{ diagrams for "left" and "right" cantilevers LC and CR}}{\sum \frac{1}{I} \Delta s}$$

$$= - \frac{\sum \frac{M}{I} \Delta s}{\sum \frac{1}{I} \Delta s} \text{(area of "right" cantilever is zero).}$$

$$V_o = + \frac{\bar{x} \text{ times area of "left" and "right" } M/I \text{ diagrams}}{\sum \frac{x^2}{I} \Delta s}$$

$$= + \frac{\sum \frac{Mx}{I} \Delta s}{\sum \frac{x^2}{I} \Delta s}$$

233

8.8

$$H = +\frac{\bar{y} \text{ times area of "left" and "right" } M/I \text{ diagrams}}{\sum \frac{y^2}{I}\Delta s}$$

$$= +\frac{\sum \frac{My}{I}\Delta s}{\sum \frac{y^2}{I}\Delta s}$$

Substituting the values obtained in the tabular summations we have

Unit load at	M/I	$1/I$	M_o ft.-tons	Mx/I	x^2/I	V_o (tons)	My/I	y^2/I	H (tons)
1	−130·95	38·40	3·410	1494·64	3289·30	0·455	164·46	124·12	1·325
3	−70·18	,,	1·828	920·58	,,	0·280	138·51	,,	1·116
5	−29·88	,,	0·778	448·30	,,	0·136	87·40	,,	0·704
7	−8·48	,,	0·221	144·88	,,	0·044	35·42	,,	0·285

Positive signs indicate that M_o, V_o, and H act in the directions assumed.

Influence Line for Bending Moment at the Crown.

From these three influence lines others may be built up. For example, the influence line for the moment at the crown of the arch follows the curve

$$M_o - (8 - 6{\cdot}29)H = M_o - 1{\cdot}71H$$

and is shown in (e).

Influence Line for Bending Moment at the Abutment.

These four influence lines are all symmetrical about the crown, and only half need be drawn. Others are unsymmetrical about the crown. For example, the influence line for M_L can be found by taking moments about L of the force and moments on the half rib.

When the unit load is on the left-hand half of the arch

$$M_L = M_o + 20V_o + 6{\cdot}29H - a$$

where a is the horizontal distance of the unit load from L.

When the unit load is on the right-hand half of the arch

$$M_L = M_o - 20V_o + 6{\cdot}29H,$$

since V_o now acts upwards on the left half of the arch.

These expressions show that there is a negative fixing moment at L when the unit load is on the left half of the arch, and a positive fixing moment when the unit load is on the right half.

234

CHAPTER 9

PIN-JOINTED STRUCTURES WITH REDUNDANT MEMBERS

The most important statically indeterminate structures are those continuous beams and frames in which stresses are imposed chiefly by bending. The increasing use of continuity in bridges and framed structures (welded construction and reinforced concrete) make the methods of the earlier chapters of importance to the structural engineer.

However, pin-jointed frames may be statically indeterminate by reason of the presence of one or more redundant members (Fig. 1.0). In such frames the members are in direct tension or compression and do not, theoretically, suffer any bending stress.

The chief difficulty experienced by the average student in the work which follows is in making rapid evaluation of the forces in the members of ordinary statically determinate frames. It is assumed, therefore, that the student is thoroughly familiar with the construction of simple stress diagrams or with the Method of Sections. These methods should be revised before proceeding further.

Procedure

The methods used in the solution of statically indeterminate pin-jointed frames are either the normal Strain Energy method (Chapter 3) or a derived Strain Energy method similar to that used for Arches (Chapter 4).

A corollary of Castigliano's Theorem (Chapter 3) is that the work done in stressing a structure under a given system of loads is the least possible consistent with the maintenance of equilibrium. The differential coefficient of the resilience or work done, with respect to one of the forces on the structure, is thus equal to zero. This "Principle of Least Work" (or "Minimum Resilience") was used in Chapter 3 to find the value of H in portal frames.

The work done (or resilience) in a member of length l and cross-sectional area A under a direct load P is $P^2l/2AE$, and the work done in the whole structure $\sum \dfrac{P^2l}{2AE}$, where P represents the force in each individual member caused by the external system of loading.

235

The differential coefficient of the work done, with respect to the force in one of the members (S), is then $\dfrac{\partial \overline{U}}{\partial S} = \sum \dfrac{Pl}{AE} \cdot \dfrac{\partial P}{\partial S}$, and this is equal to zero unless some circumstance extraneous to the loading system, such as a temperature change or lack of fit, forcibly alters the lengths l relative to each other.

FRAMES WITH ONE AND TWO REDUNDANCIES

9.1 Statically determinate or "perfect" pin-jointed frames are composed of triangles, and the forces in the members of such frames can be determined by the simple methods of statics. The "type" or simplest example of a redundant pin-jointed frame is formed by adding another member to a frame consisting of two triangles. The solution of such a frame is typical of the solution of all pin-jointed redundant frames (Fig. (a)).

The procedure is as follows:

(1) Number all the members. This is more convenient than lettering for frames of this kind.

(2) Balance the external forces, determining the magnitude and direction of the reactions. This may often be done most rapidly by means of a link or funicular polygon.

(3) Choose one member as the redundant member. It is immaterial which member is selected, but one of the diagonals is a convenient choice. Let (6) be the redundant member, the load in which is assumed to be of an unknown positive value S.

(4) Determine the loads (P) in all the members, including the redundant member. This is usually best done in two stages:

    (a) Forces due to the external loading;

    (b) Forces due to the unknown load S.

The force in the redundant member is, of course, merely $+S$. Column 2 of Table **9.1** shows these results.

(5) Determine the values of $\partial P/\partial S$ (Column 3).

(6) Sum $\dfrac{Pl}{AE} \cdot \dfrac{\partial P}{\partial S}$ for all members, including the redundant member, and equate to zero, thus determining the force S.

(7) Substitute the value of S in column 2 and determine the forces in all the members (column 7).

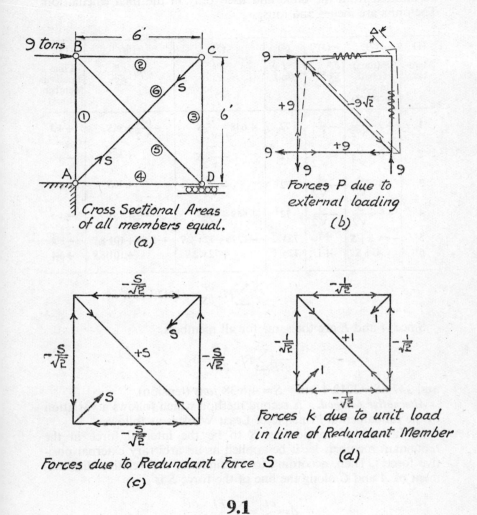

9 tons

6'

B

②

C

S

⑥

①

6'

③

S

⑤

A

S

④

D

Cross Sectional Areas
of all members equal.

(a)

9

Δ

+9

$-9\sqrt{2}$

9

+9

9

9

9

Forces P due to
external loading

(b)

$-\dfrac{S}{\sqrt{2}}$

S

$-\dfrac{S}{\sqrt{2}}$

+S

$-\dfrac{S}{\sqrt{2}}$

S

$-\dfrac{S}{\sqrt{2}}$

Forces due to Redundant Force S

(c)

$-\dfrac{1}{\sqrt{2}}$

1

$-\dfrac{1}{\sqrt{2}}$

+1

$-\dfrac{1}{\sqrt{2}}$

1

$-\dfrac{1}{\sqrt{2}}$

Forces k due to unit load
in line of Redundant Member

(d)

9.1

9.1

Any quantity (such as E), which is common to all members, may be omitted from the table and used only in the final calculation. The units are inches and tons.

(1) Member	(2) P (tons) (Load in each member)	(3) $\frac{\partial P}{\partial S}=k$	(4) l (in.)	(5) Pl	(6) $Pl \cdot \frac{\partial P}{\partial S}$	(7) Final Loads in Members (tons)
1	$+9-\dfrac{S}{\sqrt 2}$	$-\dfrac{1}{\sqrt 2}$	72	$+648-\dfrac{72}{\sqrt 2}S$	$-\dfrac{648}{\sqrt 2}+36S$	$+4\cdot5$
2	$0-\dfrac{S}{\sqrt 2}$	$-\dfrac{1}{\sqrt 2}$	72	$-\dfrac{72}{\sqrt 2}S$	$+36S$	$-4\cdot5$
3	$0-\dfrac{S}{\sqrt 2}$	$-\dfrac{1}{\sqrt 2}$	72	$-\dfrac{72}{\sqrt 2}S$	$+36S$	$-4\cdot5$
4	$+9-\dfrac{S}{\sqrt 2}$	$-\dfrac{1}{\sqrt 2}$	72	$+648-\dfrac{72}{\sqrt 2}S$	$-\dfrac{648}{\sqrt 2}+36S$	$+4\cdot5$
5	$-9\sqrt2+S$	$+1$	$72\sqrt2$	$-1296+72\sqrt2S$	$-1296+101\cdot8S$	$-6\cdot4$
6	$0+S$	$+1$	$72\sqrt2$	$+72\sqrt2S$	$+101\cdot8S$	$+6\cdot4$

$$\sum Pl \cdot \frac{\partial P}{\partial S}=-2212\cdot4+347\cdot6S$$

Since A and E are the same for all members,

$$\frac{1}{AE}\sum Pl \cdot \frac{\partial P}{\partial S}=0$$

and $347\cdot6S=2212\cdot4$ or $S=+6\cdot38$ *tons* (tension).

Alternative Method. A second method which follows a variation of the fundamental Principle of Least Work is as follows:

(1) Instead of considering S to be the internal force in the redundant member, let it be applied as an arbitrary external positive force. Then, according to Castigliano, the relative displacement of A and C along the line of the force S is

$$\varDelta=\frac{\partial \bar U}{\partial S}=\sum \frac{Pl}{AE}\cdot\frac{\partial P}{\partial S}$$

where the summation is for the five members (1) to (5) only, the member (6) having been removed.

(2) But $\partial P/\partial S$ represents the forces in all the members induced by a unit load in place of the force S. This can be realised from column 3 of Table **9.1,** and from **(d).** Let these values be $=k$.

238

(3) Then $\Delta = Pkl/AE$ and S, which was arbitrary, can be reduced to zero so that $\Delta =$ the relative displacement of A and C along the diagonal, due to the external loading only (**b**). P is now represented by column 2, with S made equal to zero. In other words, P represents the forces in the members due to the imposed loading when the redundant member has been removed.

(4) If this displacement Δ is to be eliminated (as it is in Fig. (**a**)) an inward force along AC must be employed. Find the relative displacement of A and C due to a unit load acting along the line of the diagonal. This displacement will be $\sum \dfrac{k \cdot kl}{AE}$, k taking the place of P in the expression given in (3).

(5) Thus the number of "unit loads" or the value of S required to bring A and C back to their original positions is given by
Outward displacement due to loads
$\qquad$ —inward displacement due to $S = 0$

$$\sum \frac{Pkl}{AE} + S \sum \frac{k^2 l}{AE} = 0$$

the summation being made for members (1) to (5) only.

(6) However, in actual fact the deflection Δ is not entirely eliminated in the frame of Fig. (**a**), for the force S is not applied externally, but by an extensible member. Member (6), then, extends through a length Sl_6/A_6E. Thus, when S is applied by means of the redundant member, we have
Outward displacement due to loads
$\qquad$ —inward displacement due to $S =$ extension of (6)

$$\sum \frac{Pkl}{AE} + S \sum \frac{k^2 l}{AE} = -\frac{Sl_6}{A_6E}$$

(7) But k for member (6) $= 1$, so we may write

$$\sum \frac{Pkl}{AE} + S \sum \frac{k^2 l}{AE} = -\frac{k_6^2 l_6}{A_6E}$$

$$\therefore S = -\frac{\sum \dfrac{Pkl}{AE}}{\sum \dfrac{k^2 l}{AE}}$$

the numerator being summed for all members *except* the redundant member, and the denominator being summed for all members *including* the redundant member.

After reading Problem **9.2** rewrite the table according to this method.

9.2 In (a) AB is a tie bar. The members in tension have a cross-sectional area of 1 sq. in., and the members in compression have a cross-sectional area of 2 sq. in.

(1) Remove the redundant member AB and find the forces P in all the members due to the external loading only (Fig. (b) and column 2). Use the Method of Sections or a stress diagram.

(2) Remove the 5-ton load and apply a positive load of 1 ton along the line of action of member (6). The loads in the members of the frame due to the unit load are the values k of column 3.

(3) Sum Pkl/A for all members *except AB* and k^2l/A for all members *including AB*, and complete the table.

Solve this problem in the way shown in Problem **9.1**, using an unknown force S in the redundant member.

(1)	(2)	(3)	(4)	(5)	(6)	(7)	(8)
							Final Loads in Members $P+Sk$
Member	P	k	l	A	Pkl/A	k^2l/A	
	(tons)	(tons)	(in.)	(oz. in.)			(tons)
1	−6·25	+0·63	60	2	−117	+ 18·7	−4·51
2	−6·25	+0·63	60	2	−117	+ 18·7	−4·51
3	+2·50	−0·75	24	1	− 45	+ 13·5	+0·42
4	+5·16	−1·55	50	1	−396	+119·0	+0·85
5	+5·16	−1·55	50	1	−396	+119·0	−9·47
					$\sum \dfrac{Pkl}{A} = -1071$		
6	—	+1·00	96	1	—	+ 96·0	+2·78
						$\sum \dfrac{k^2l}{A} = + 385$	

$$S = -\frac{\sum \dfrac{Pkl}{AE}}{\sum \dfrac{k^2l}{AE}} = \frac{1071}{385} = +2 \cdot 78 \ tons \ \text{(tension)}.$$

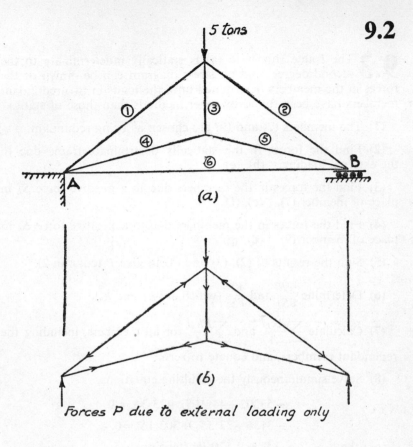

(a)

(b)

Forces P due to external loading only

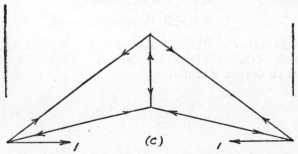

(c)

Forces k due to unit load in line of
redundant member

9.3 The frame shown in (a) is statically indeterminate to the second degree, and no stress diagram can be drawn or the forces in the members determined until the loads in two redundant members have been found by other methods than those of statics.

(1) The members (7) and (9) are chosen as being redundant.

(2) Find the forces in the statically determinate frame due to the external loading. **(b), (e)**

(3) Find the forces in the members due to a positive force S_L in place of member (7). **(c), (f)**

(4) Find the forces in the members due to a positive force S_u in place of member (9). **(d), (g)**

(5) Sum the results of (2), (3) and (4) to give P (column 2).

(6) Determine $\dfrac{\partial P}{\partial S_L}$ and $\dfrac{\partial P}{\partial S_u}$ (which are k_L and k_u).

(7) Calculate $\sum \dfrac{Pk_L l}{A}$ and $\sum \dfrac{Pk_u l}{A}$ for all members, including the redundant members, and equate to zero.

(8) Solve simultaneously the resulting equations,

$$-53202 + 445 \cdot 1 S_L + 23 \cdot 3 S_u = 0,$$
$$+3136 + 23 \cdot 3 S_L + 303 \cdot 1 S_u = 0.$$

From these, $S_L = +120$ *lb.* (tension),

$S_u = -20$ *lb.* (compression).

(9) Substitute the values of S_L and S_u in column 2 and find the final values of the loads in all the members. The final force diagram can now be drawn, if required.

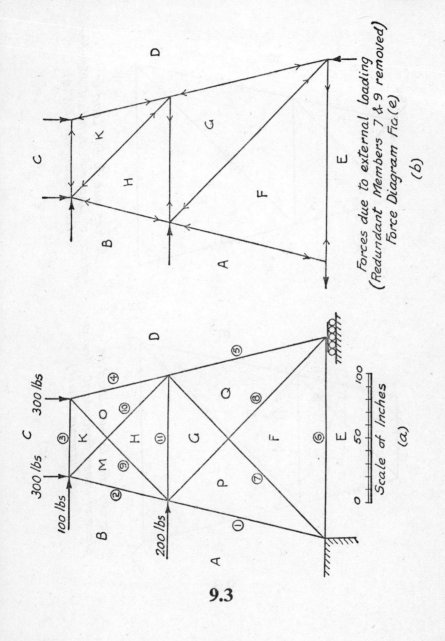

Forces due to external loading
(Redundant Members 7 & 9 removed)
Force Diagram Fig.(e)

(b)

300 lbs

300 lbs C

300 lbs

100 lbs

B

200 lbs

A

Scale of Inches

(a)

9.3

9.3

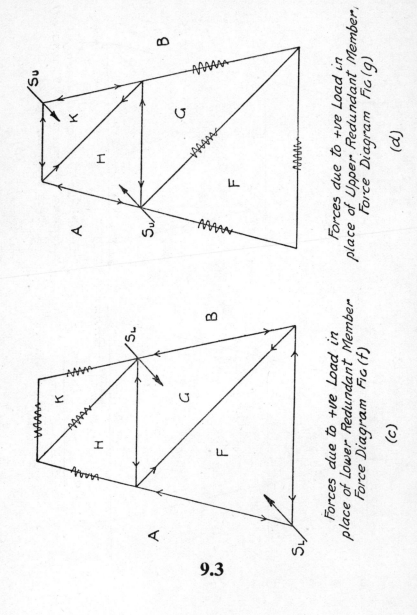

Forces due to +ve Load in
place of Upper Redundant Member,
Force Diagram Fig (g)

(d)

Forces due to +ve Load in
place of Lower Redundant Member
Force Diagram Fig (f)

(c)

9.3

244

(1) Member	(2) P (lb.)		(3) $\partial P/\partial S_L$ (k_L)	(4) $\partial P/\partial S_u$ (k_u)	(5) l in.	(6) A sq. in.	(7) l/A
1	-7	$-0.74S_L$	-0.74	—	130·5	3	43·5
2	-224	$-0.74S_u$	—	-0.74	80·5	2	40·3
3	-70	$-0.89S_u$	—	-0.89	60·0	2	30·0
4	-308	$-0.74S_u$	—	-0.74	80·5	2	40·3
5	-392	$-0\;74S_L$	-0.74	—	130·5	3	43·5
6	$+303$	$-0.54S_L$	-0.54	—	158·0	2	79·0
7		$+S_L$	$+1$	—	168·0	1	168·0
8	-300	$+S_l$	$+1$	—	168·0	1	168·0
9		$+S_u$	—	$+1$	110·5	1	110·5
10	-117	$+S_u$	—	$+1$	110·5	1	110·5
11	$+63$	$-0.89S_L\ -0.54S_u$	-0.89	-0.54	97·0	2	48·5

(8) Pl/A		(9) $Pk_L l/A$			(10) $Pk_u l/A$		
-305	$-32.2S_L$	$+226$	$+23.8S_L$		—		
$-9,016$	$-29.8S_u$	—			$+6,672$	$+22.1S_u$	
$-2,100$	$-26.7S_u$	—			$+1,869$	$+23.8S_u$	
$-12,397$	$-29.8S_u$	—			$+9,174$	$+22.1S_u$	
$-17,052$	$-32.2S_L$	$+12,618$	$+23.8S_L$		—		
$+23,937$	$-42.7S_L$	$-12,926$	$+23.1S_L$		—		
	$+168.0S_L$		$+168.0S_L$		—		
$-50,400$	$+168.0S_L$	$-50,400$	$+168.0S_L$		—		
	$+110.5S_u$	—				$+110.5S_u$	
$-12,929$	$+110.5S_u$	—			$-12,929$	$+110.5S_u$	
$+3,056$	$-43.2S_L\ -26.2S_u$	$-2,720$	$+38.4S_L$	$+23.3S_u$	$-1,650$	$+23.3S_L$	$+14.1S_u$
Sum:		$-53,202$	$+445.1S_L$	$+23.3S_u$	$+3,136$	$+23.3S_L$	$+303.1S_u$

245

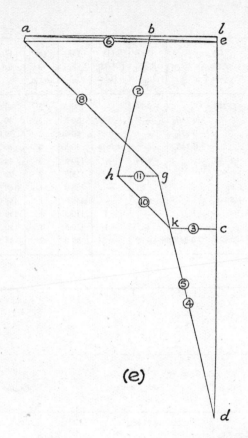

(e)

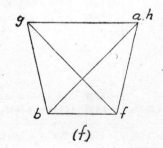

(f)

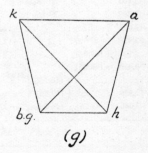

(g)

Revision Exercises

Solve each of these problems by at least two methods:

The loading on a rectangular two-hinged frame is 1 ton per ft. run, acting horizontally against the whole length of the leg AB. If $I_{AB}:I_{BC}:I_{CD}=1:1\frac{1}{2}:1\frac{3}{4}$, drawn the B.M.D. and deflected form. $AB=CD=10$ ft. : $BC=7$ ft.
$$(M_B=17\cdot2 \text{ ft.-tons.} \quad M_C=32\cdot8 \text{ ft.-tons.})$$

A triangular two-hinged frame of constant moment of inertia is hinged at A and C and has a rigid joint at B. Both legs slope at 45° to the horizontal, and the distance between A and C is 28 ft. $3\frac{1}{2}$ in. A uniformly distributed horizontal wind load of $\frac{1}{2}$ ton per foot of height acts over the whole vertical height of the structure. Draw the B.M.D. and deflected form. $(M_B=24 \text{ ft.-tons.})$

A beam AB is 12 ft. long and is fixed horizontally at A. At B there is a right-angled stiff joint leading to an extension (BC) 2 ft. long, projecting vertically downwards from B. At C a pull of 2 tons is exerted horizontally in a direction away from A. B is prevented from rising, and is kept at the level of A by a knife-edge reaction. Solve for B.M.D. $(M_A=2 \text{ ft.-tons.})$

Both legs of a rectangular two-hinged frame are subjected to a hydrostatic pressure varying from zero at half the height of the leg to a maximum at the base. The height of the frame $(AB=CD)$ is 15 ft. and $BC=5$ ft. $I_{AB}:I_{BC}:I_{CD}=1:2:1$. Draw the B.M.D. and deflected form. $(M_B=1620 \text{ ft.-lb.})$

A beam AB is built in horizontally at both ends, and carries a U.D.L. of $\frac{3}{4}$ ton per foot over the whole span of 24 ft. At 16 ft. from A the beam also carries a concentrated load of 12 tons. $(M_A=-57\cdot3 \text{ ft.-tons.})$

A vertical cantilever (the wall of a tank holding water) is $l+a$ ft. long. The top of the wall is held in position at a point vertically above the base, but not fixed in direction. The base of the wall is fixed vertically. The surface of the water is a ft. from the top of the wall, and the water is l ft. deep. Show that the horizontal reaction holding the top of the wall in position is $7\cdot8l^4\{a+0\cdot8l/(a+l)^3\}$.

In a two-hinged rectangular frame $AB=6$ ft. ; $BC=8$ ft. ; $CD=10$ ft. $I_{AB}:I_{BC}:I_{CD}=1:2:1\frac{1}{4}$. The loading consists of a vertical concentrated load of 5 tons on the beam BC at 3 ft. from B. Draw the B.M.D. and deflected form.
$(H=0\cdot25 \text{ tons.})$

A frame of the type shown in 3.11 is loaded by horizontal loads acting inwards on the legs AB and EF. The loads are uniformly distributed over the whole height of the legs at an intensity of $\frac{1}{2}$ ton per foot of height. The moment of inertia of the cross-section is constant throughout the frame. $AB=CD=EF=10$ ft ; $BC=CE=8$ ft. Draw B.M.D. and deflected form. $(M_B=3\cdot9 \text{ ft.-tons.})$

247

9.4 When a member of a statically determinate pin-jointed frame is slightly too long or too short, the difference has no effect on the forces in the various members. If, however, one of the members of a redundant frame is not of the exact length required and is forced into place, forces are thereby induced in all the other members of the frame.

In the simple frame of (a) BC is found to be 0·05 in. too long and is forced into place. A compression force is thus induced in the member BC (force T) and other forces, consistent with the magnitude of T, are induced in the other members. If $E=13,000$ tons per sq. in., the problem is to determine the forces in the members due to this lack of fit.

Remove the redundant member (BC) and replace it by a unit force (b). Then $\sum \frac{k^2 l}{AE}$ (for all members *except BC*) is the relative displacement of B and C due to the unit loading.

But the load in BC is T. The displacement of B relative to C, due to a force T, is $T \sum \frac{k^2 l}{AE}$ (T times the displacement due to 1 ton).

If the member BC were completely incompressible, this relative displacement of B and C from their original positions would be equal to the extra length of the member BC (0·05 in.). However, the member BC is of steel, and shortens under the compressive load T. The final relative displacement of B and C is thus

$$0·05 \ in. - T l_2 / A_2 E.$$

As found above, the relative displacement of B and C is $T \sum \frac{k^2 l}{AE}$ for all members by BC. We thus have

$$T \sum \frac{k^2 l}{AE} = 0·05 \ in. - \frac{T l_2}{A_2 E}.$$

Since the value of the "k" force in BC is unity, $T l_2 / A_2 E$ may be written $T k_2^2 l_2 / A_2 E$. Substituting in the above equation we have the final expression

$$T \sum \frac{k^2 l}{AE} = 0·05 \ in.$$

where $\sum \frac{k^2 l}{AE}$ is obtained for all members, including BC.

248

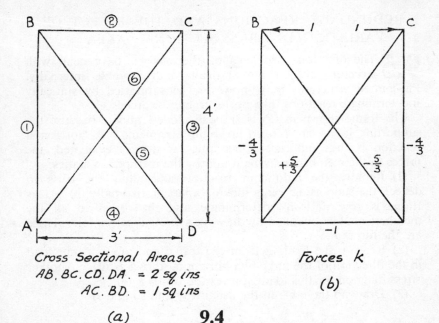

Cross Sectional Areas
AB, BC, CD, DA. = 2 sq ins
AC. BD. = 1 sq ins

(a)

Forces k

(b)

9.4

Member	k (tons)	l (in.)	A (sq. in.)	k^2l/A	Final Forces ($k \times 1 \cdot 43$)
1	−4/3	48	2	42·67	−1·91
2	−1	36	2	18·00	−1·43
3	−4/3	48	2	42·67	−1·91
4	−1	36	2	18·00	−1·43
5	+5/3	60	1	166·67	+2·48
6	+5/3	60	1	166·67	+2·48

Sum: 454·67

$$T = \frac{0 \cdot 05}{\sum \frac{k^2l}{AE}} = \frac{0 \cdot 05 \times 13{,}000}{454 \cdot 67}$$

$= 1 \cdot 43$ *tons* (in the direction assumed: compressive).

From this, the forces in the other members can be determined and the values inserted in the last column.

Lack of fit may be caused by a relative difference in temperature between the members. The change in length of the redundant member which would be caused by the *difference* in temperature between the redundant member and the rest, is the figure which takes the place of 0.05 in. in the above problem.

REDUNDANT REACTIONS AND THE EFFECT OF VARIATION IN CROSS-SECTIONAL AREA

9.5 The first four problems of this chapter have dealt with structures which are statically indeterminate internally. Pin-jointed structures may, however, be supported by statically indeterminate reactions, like portal frames or arches.

The frame shown in (a) is statically indeterminate because the supporting joints are hinged to rigid abutments. A horizontal reaction is thus induced. Once this reaction is evaluated, the forces in the members may be found by the methods of statics.

(1) Remove the redundant reaction and allow the truss to deflect, the supporting joints moving apart, horizontally (b). This frame is now statically determinate and the forces in all the members may be obtained by drawing a stress diagram (c). These are the forces P.

(2) Remove the loading from the frame and apply a unit load in the direction of the unknown horizontal thrust H (d). Another stress diagram for this loading gives the forces k (e).

(3) Draw up the table in the usual way and so find H.

$$H = -\frac{\sum \dfrac{Pkl}{AE}}{\sum \dfrac{k^2 l}{AE}} = \frac{645 \cdot 5}{194 \cdot 2} = 3 \cdot 3 \ tons \ \text{(see Problem 9.1).}$$

(4) Complete the last column in order to find the final forces in all the members when the structure is loaded as in (a). The final forces are

$$P + 3 \cdot 3k.$$

Member	P (tons)	k (tons)	k^2	l (ft.)	A (sq.in.)	$\dfrac{Pkl}{A}$	$\dfrac{k^2 l}{A}$	Final Forces in Members $P+3\cdot3k$ (tons)
1	−17·4	+2·8	+7·84	10·00	4	−121·8	+19·6	−17·4+ 9·2=−8·2
2	−13·7	+3·8	+14·44	3·66	4	−47·6	+13·2	−13·7+12·6=−1·1
3	−10·0	+2·8	+7·84	10·00	4	−70·0	+19·6	−10·0+ 9·2=−0·8
4	+7·1	−3·4	+11·56	12·25	2	−147·9	+70·9	+ 7·1−11·2=−4·1
5	+12·4	−3·4	+11·56	12·25	2	−258·2	+70·9	+12·4−11·2=+1·2

Sum: −645·5   +194·2

Note.—The conversion of l to inches has not been made, since the same factor would be required in numerator and denominator alike.

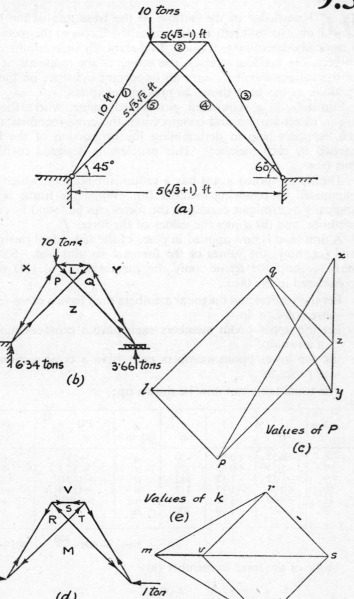

10 tons

5(√3−1) ft

②

①

10 ft

5√3√2 ft

⑤

④

③

45°

60°

5(√3+1) ft

(a)

10 Tons

X

L

P

Q

Y

Z

6·34 tons

3·66 tons

(b)

x

q

z

l

y

p

Values of P

(c)

V

S

R

T

M

i ton

i ton

(d)

Values of k

(e)

r

m

v

s

t

9.6 Knowledge of the outline of the truss and of the loading on it is sufficient to determine the forces in the members of a statically determinate frame. In a statically indeterminate frame subjected to bending stresses, the values of the moments of inertia of the various members have an important influence on the forces induced, as has been shown in previous chapters.

Similarly, in a redundant pin-jointed frame, where the forces are in direct tension and compression, the cross-sectional areas of the members help in determining the proportion of the loading carried by each member. This problem is designed to illustrate this fact.

The frame shown in (a) has a redundant reaction which can be eliminated by removing member (8). When the frame is in this statically determinate condition the forces can be found by ordinary methods, and (b) shows the values of the forces P.

A unit load is now applied in place of the redundant member (8), and (c) shows the values of the forces k so obtained. Since k is zero for half the frame, only the members (6) to (11) need be considered in the table.

Let the vertical and diagonal members each have a cross-sectional area of a sq. in.
Let the upper boom members each have a cross-sectional area of na sq. in.
Let the lower boom members each have a cross-sectional area of ra sq. in.

The usual table can now be drawn up:

Member	$\dfrac{P}{W}$	k	l (in.	A (sq. in.)	$\dfrac{Pkl}{A} \times \dfrac{a}{W}$	$\dfrac{k^2 l}{A} \times a$
6	$-0{\cdot}50$	$+0{\cdot}50$	50	a	$-12{\cdot}5$	$+12{\cdot}5$
7	$-1{\cdot}50$	$+0{\cdot}50$	50	a	$-37{\cdot}5$	$+12{\cdot}5$
9	$+1{\cdot}12$	$-1{\cdot}12$	112	a	$-140{\cdot}0$	$+140{\cdot}0$
10	$+3{\cdot}36$	$-1{\cdot}12$	112	a	$-419{\cdot}0$	$+140{\cdot}0$
11	$-4{\cdot}00$	$+1{\cdot}00$	100	ra	$-400/r$	$+100/r$
8	—	$+1{\cdot}00$	100	na	—	$+100/n$

$$\text{Sum:} \quad -609 - \frac{400}{r} \qquad\qquad 305 + \frac{100}{r} + \frac{100}{n}$$

Value of the force in member (8)

$$F_8 = -\frac{\sum \dfrac{Pkl}{AE}}{\sum \dfrac{k^2 l}{AE}} = \frac{6{\cdot}09 + \dfrac{4}{r}}{3{\cdot}05 + \dfrac{1}{r} + \dfrac{1}{n}}\ W\ lb.\ \text{tension.}$$

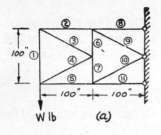

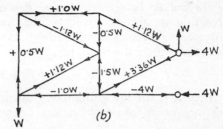

Forces P due to external loading
when redundant member removed

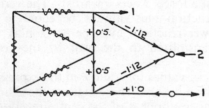

Forces k due to unit load
in place of redundant member

(c)

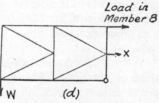

Load in
Member 8

Determination of the
Horizontal Component
of the Centre Reaction

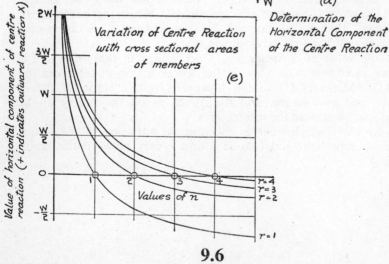

Variation of Centre Reaction
with cross sectional areas
of members

(e)

9.6

To find the Value of the Centre Reaction. The value of the upper reaction is always equal to the force in member (8). Taking moments of the external loading about the lower hinge, where the bending moment is zero, (d)

$$-2W+F_8+X/2=0$$

where X is the horizontal component of the centre reaction. The expression, after simplification, becomes

$$X=\frac{4(r-n)}{3\cdot05nr+r+n}W \ lb.$$

If this expression is positive, the force X acts towards the right (outwards), and if negative, the force X acts inwards. The vertical component of the centre reaction must always be equal to W, since both the upper and lower reactions must be horizontal, and the whole vertical shear transmitted to the wall by the centre hinge.

The substitution of numerical values for r and n in the expression for X results in the curves shown in (e). These indicate the changes which occur in the horizontal component of the centre reaction as the cross-sectional areas of the upper and lower booms vary in value.

When the cross-sectional area of the upper boom is greater than that of the lower boom, the reaction acts inwards. When the cross-sectional area of the upper boom is less than that of the lower boom, the reaction acts outwards. When the boom areas are equal, the special case of $X=0$ is obtained. Such special cases are shown by ringed points in (e).

As the value of X varies with fluctuations in that of n and r similar variations take place in the forces acting in members (6) to (11). Members (1) to (5), however, form a statically deter-minate cantilever frame, and the forces which they carry in (b) never alter, whatever the change in X.

As an exercise, plot curves showing the variation of the forces in the members (9) and (11) as n and r vary between the limits 1 and 4.

254

THE TRUSSED BEAM

9.7 The earlier chapters have dealt with structures whose members are subjected to bending, and in this chapter the problems have been concerned with direct stress. Occasionally a structure comprises members subjected to bending and members subjected to direct stress. The best known is the "trussed beam," which is used, for example, in the construction of railway passenger coaches.

Fig. (a) shows a timber beam (12 in. × 12 in.) strengthened by means of two cast-iron struts each of 3 sq. in. cross-sectional area (members (2) and (3)), and two steel tie bars each 1 in. diameter (members (4), (5), and (6)). The beam carries a load of 1000 lb. per ft.

Members (2) to (6) carry direct stress only, while member (1) carries both direct stress and bending stress.

The loading and construction are both symmetrical, and thus (2) and (3) carry the same load. Let these members be removed and replaced by a negative load S (b).

Strain Energy due to Direct Stress. The first table is completed by considering only the direct stress caused by the forces S, as if all the joints were pin joints.

$$\frac{\partial \bar{U}}{\partial S} = \sum \frac{Pl}{AE} \cdot \frac{\partial P}{\partial S} = \sum \frac{Pkl}{AE}$$

for all members.

It should be noted that, in this structure, three different materials are used. The value of Young's Modulus is not constant throughout, but must be brought into the table as one of the variables.

Forces in the members can be found by drawing the triangle of forces for joint A (b). Member (1) carries the horizontal component of the force in member (4).

Member	P (lb.)	$\partial P/\partial S$ (k)	l (in.)	A (sq. in.)	$E/10^6$ (lb./sq. in.)	$\dfrac{Pkl}{AE} \times 10^6$
1	$-4.8S$	-4.8	396	144	1·5	42·24 S
2	$-S$	-1.0	24	3	20	0·40 S
3	$-S$	-1.0	24	3	20	0·40 S
4	$+4.9S$	$+4.9$	147	1·57	30	74·94 S
5	$+4.8S$	$+4.8$	108	1·57	30	52·82 S
6	$+4.9S$	$+4.9$	147	1·57	30	74·94 S

Sum: 245·74 S

255

9.7

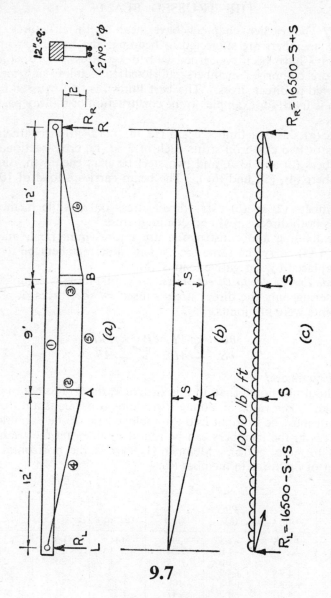

9.7

256

Displacement in the direction of S, due to direct stress, is

$$\frac{\partial \overline{U}}{\partial S} = \sum \frac{Pkl}{AE} = 0.000246S.$$

Strain Energy due to Bending Stress. There is resilience or strain energy stored in member (1) due to the bending stress which is imposed by the uniformly distributed load of 1000 lb. per ft. The forces acting on the beam are shown in (c).

Of the inclined force (member (4)) only the vertical component need be considered, since the horizontal component has no effect on the bending moment in member (1), but merely applies a direct compressive stress in that member. The nett vertical force at L, therefore, is $16,500 - S$ lb.

EI (member (1)) $= 1.5 \times 10^6 \times 144 \times \frac{1}{12} = 18 \times 10^6$ *lb.-ft.*^{2}

Member	Bending Moment	$\partial M/\partial S$	Limits
LA and RB . .	$(16,500-S)x-500x^2$	$-x$	0–12
AB . .	$(16,500-S)x-500x^2+S(x-12)$	-12	12–21

$$\frac{2}{EI}\int_0^{12}\{(16,500-S)x-500x^2\}(-x)dx = \frac{1}{18\times 10^6}(-13.824\times 10^6 + 1152S)$$

$$\frac{1}{EI}\int_{12}^{21}(16,500x-500x^2-12S)(-12)dx$$

$$= \frac{1}{18\times 10^6}(-14.337\times 10^6 + 1296S)$$

$$\text{Sum} = \frac{\partial \overline{U}}{\partial S} = -1.5645 + 0.000136S \ ft.$$

$$\frac{\partial \overline{U}}{\partial S} = -18.774 + 0.001632S \ in.$$

Value of S. Summing $\frac{\partial \overline{U}}{\partial S}$ for direct stress and $\frac{\partial \overline{U}}{\partial S}$ for bending stress, and equating to zero, we have

$$-18.774 + 0.001878S = 0,$$

whence $$S = 10,000 \ lb.$$

257

PICTORIAL INDEX

— indicates varying moment of inertia of cross section.

☐ indicates frame sways laterally under load.

○ indicates there is settlement of supports.

TYPE	PROBLEM

Beams and Cantilevers

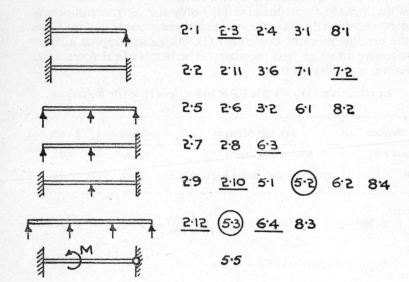

	2·1 2·3 2·4 3·1 8·1
	2·2 2·11 3·6 7·1 7·2
	2·5 2·6 3·2 6·1 8·2
	2·7 2·8 6·3
	2·9 2·10 5·1 (5·2) 6·2 8·4
	2·12 (5·3) 6·4 8·3
	5·5

PORTAL FRAMES

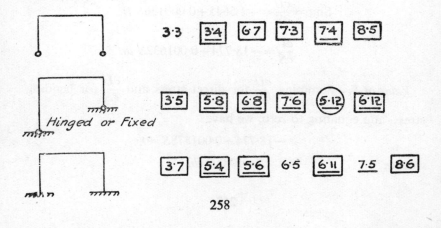

	3·3 [3·4] [6·7] [7·3] [7·4] [8·5]
Hinged or Fixed	[3·5] [5·8] [6·8] [7·6] ([5·12]) [6·12]
	[3·7] [5·4] [5·6] 6·5 [6·11] 7·5 [8·6]

258

Bridge and Building Frames

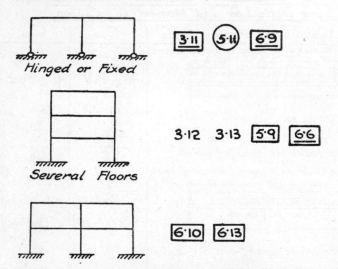

Hinged or Fixed   3·11   (5·11)   6·9

Several Floors   3·12   3·13   5·9   6·6

                     6·10   6·13

Frames with Inclined Members

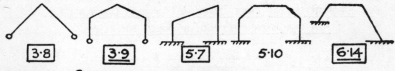

3·8   3·9   5·7   5·10   6·14

Arches   Segmental and Parabolic

3·10   3·14   3·15   8·7   8·8   and Chapter . 4 .

Pin-Jointed Frames with Redundant Members
Chapter . 9 .

APPENDIX

FIXED-END MOMENTS *for use with methods of Slope Deflection and Moment Distribution*

TYPE LOADING AND SPECIAL CASES	C_{AB}	C_{BA}
ANY NUMBER OF CONCENTRATED LOADS 	$\frac{1}{l^2}\sum Pab^2$	$\frac{1}{l^2}\sum Pba^2$
Single Load $a = b$	$\frac{Pl}{8}$	$\frac{Pl}{8}$
Two loads at "third points"	$\frac{2Pl}{9}$	$\frac{2Pl}{9}$
Three loads at "quarter points"	$\frac{15Pl}{48}$	$\frac{15Pl}{48}$
UNIFORM LOAD OVER ANY PORTION OF SPAN	$\frac{w}{12l^2}\left[d^3(4l-3d)-b^3(4l-3b)\right]$	$\frac{w}{12l^2}\left[a^3(4l-3a)-c^3(4l-3c)\right]$
U.D.L. over whole span $c=0 \ b=0 \ a=d=l$	$\frac{wl^2}{12}$	$\frac{wl^2}{12}$
U.D.L. over half span from A to centre $c=0 \ a=b=\frac{l}{2} \ d=l$	$\frac{11wl^2}{192}$	$\frac{5wl^2}{192}$

PROPERTIES OF FREE BENDING MOMENT DIAGRAM
FOR USE WITH METHOD OF AREA MOMENTS.

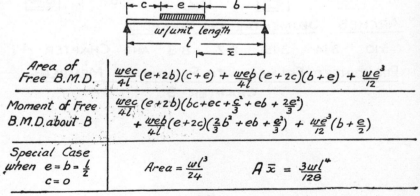

Area of Free B.M.D.	$\frac{wec}{4l}(e+2b)(c+e) + \frac{web}{4l}(e+2c)(b+e) + \frac{we^3}{12}$
Moment of Free B.M.D. about B	$\frac{wec}{4l}(e+2b)(bc+ec+\frac{c^2}{3}+eb+\frac{2e^2}{3})$ $+ \frac{web}{4l}(e+2c)(\frac{2}{3}b^2+eb+\frac{e^2}{3}) + \frac{we^3}{12}(b+\frac{e}{2})$
Special Case when $e=b=\frac{l}{2}$ $c=0$	$Area = \frac{wl^3}{24}$ $\quad A\bar{x} = \frac{3wl^4}{128}$

260